# ERGEBNISSE DER MATHEMATIK UND IHRER GRENZGEBIETE

UNTER MITWIRKUNG DER SCHRIFTLEITUNG DES „ZENTRALBLATT FÜR MATHEMATIK“

HERAUSGEGEBEN VON

L. V. AHLFORS · R. BAER · F. L. BAUER · R. COURANT · A. DOLD
J. L. DOOB · S. EILENBERG · P. R. HALMOS · M. KNESER
T. NAKAYAMA · H. RADEMACHER · F. K. SCHMIDT
B. SEGRE · E. SPERNER

NEUE FOLGE · HEFT 28

REIHE:

# MODERNE FUNKTIONENTHEORIE

BESORGT
VON

L. V. AHLFORS

SPRINGER-VERLAG
BERLIN · GÖTTINGEN · HEIDELBERG
1960

# CLUSTER SETS

BY

KIYOSHI NOSHIRO

SPRINGER-VERLAG
BERLIN · GÖTTINGEN · HEIDELBERG
1960

ISBN-13: 978-3-540-02516-0 e-ISBN-13: 978-3-642-85928-1
DOI: 10.1007/978-3-642-85928-1

# Preface

For the first systematic investigations of the theory of cluster sets of analytic functions, we are indebted to IVERSEN [1—3] and GROSS [1—3] about forty years ago. Subsequent important contributions before 1940 were made by SEIDEL [1—2], DOOB [1—4], CARTWRIGHT [1—3] and BEURLING [1]. The investigations of SEIDEL and BEURLING gave great impetus and interest to Japanese mathematicians; beginning about 1940 some contributions were made to the theory by KUNUGUI [1—3], IRIE [1], TÔKI [1], TUMURA [1—2], KAMETANI [1—4], TSUJI [4] and NOSHIRO [1—4]. Recently, many noteworthy advances have been made by BAGEMIHL, SEIDEL, COLLINGWOOD, CARTWRIGHT, HERVÉ, LEHTO, LOHWATER, MEIER, OHTSUKA and many other mathematicians. The main purpose of this small book is to give a systematic account on the theory of cluster sets.

Chapter I is devoted to some definitions and preliminary discussions. In Chapter II, we treat extensions of classical results on cluster sets to the case of single-valued analytic functions in a general plane domain whose boundary contains a compact set of essential singularities of capacity zero; it is well-known that HÄLLSTRÖM [2] and TSUJI [7] extended independently Nevanlinna's theory of meromorphic functions to the case of a compact set of essential singularities of logarithmic capacity zero. Here, Ahlfors' theory of covering surfaces plays a fundamental rôle. Chapter III is concerned with functions meromorphic in the unit circle. We discuss here functions of class $(U)$ in Seidel's sense, boundary theorems of COLLINGWOOD-CARTWRIGHT, recent important results of BAGEMIHL-SEIDEL and COLLINGWOOD on the relation between Baire category and cluster sets, Bagemihl's results on ambiguous points, Meier's results related to Lusin-Privaloff-Plessner's theorem and results of LEHTO and VIRTANEN on meromorphic functions of bounded type and normal meromorphic functions. In Chapter IV, we deal with single-valued analytic functions on open Riemann surfaces and discuss covering properties and boundary behaviours. We state here some recent results of HEINS, KURODA, KURAMOCHI and CONSTANTINESCU-CORNEA from the view-point of cluster sets. We hope that these fragmentary treatments will contribute to the future theory of cluster sets of analytic functions on open Riemann surfaces. Appendix is devoted to cluster sets of pseudo-analytic functions. A recent paper of BEURLING-AHLFORS [1] contains a striking result from the view-point of cluster sets. We cannot apply the theory of functions of class ($U$) in Seidel's sense to the case of pseudo-analytic functions without any additional condition. We discuss to what extent results on cluster sets of analytic functions can be extended to the case of pseudo-analytic functions.

It has been my earnest desire to write a systematic account on cluster sets since some years ago. I should like to express my hearty thanks to Professor LARS V. AHLFORS for his kind recommendation to the Ergebnisse Series. I am very grateful to my colleague Professor T. KURODA and my students Mr. R. IWAHASHI, Mr. M. KISHI and Mr. M. NAKAI for their careful readings of the manuscript and for their helpful comments. It is also a pleasure to acknowledge the constant generosity and courtesy of the Springer Verlag.

November 10, 1959 KIYOSHI NOSHIRO
Nagoya

# Contents

# I. Definitions and preliminary discussions

## § 1. Definitions of Cluster Sets

**1.** Let $D$ be an arbitrary domain[1] with boundary $\Gamma$. Let $E$ be a totally disconnected closed set contained in $\Gamma$. We suppose that $w = f(z)$ is non-constant, single-valued and meromorphic in $D$. We associate with every point $z_0$ of $\Gamma$ the following sets of values.

(i) The **Cluster Set** $C_D(f, z_0)$. $\alpha \in C_D(f, z_0)$ if there exists a sequence of points $\{z_n\}$ with the following properties

$$z_n \in D\,, \quad \lim_{n\to\infty} z_n = z_0\,, \quad \lim_{n\to\infty} f(z_n) = \alpha\,. \tag{1}$$

If we denote by $\mathfrak{D}_r$ the set of values of $w = f(z)$ in the intersection $D_r$ of $D$ with a circular disc $|z - z_0| < r$, then

$$C_D(f, z_0) = \bigcap_{r>0} \overline{\mathfrak{D}}_r\,, \tag{2}$$

where $\overline{\mathfrak{D}}_r$ denotes the closure of $\mathfrak{D}_r$[2].

Evidently $C_D(f, z_0)$ is a non-empty closed set. In the particular case where $D$ is a Jordan domain bounded by a simple closed curve, $C_D(f, z_0)$ is either a single point or a continuum. However, this property does not hold in general cases[3].

Remark. Consider the special case where $z_0$ is an accessible boundary point of $D$. Then, there exists a path (simple curve) $L$ in $D$ terminating at $z_0$. Denote by $z_r$ the last point of intersection of $L$ with a circle $c$: $|z - z_0| = r$ and by $L_r$ the arc $\widehat{z_r, z_0}$ of $L$; such an arc is called a last part of $L$. The intersection $D_r$ of $D$ with $(c)$: $|z - z_0| < r$ is an open set which consists of at most an enumerably infinite number of connected components. Let $\Delta_r$ be the component which contains the last part $L_r$ of $L$. If we denote by $\mathfrak{D}_r^*$ the value set of $f(z)$ in $\Delta_r$, then $\mathfrak{D}_r^*$ is a domain and, hence, $\overline{\mathfrak{D}_r^*}$ is a continuum. Hence the set

$$C_D(f, z_0; L) = \bigcap_{r>0} \overline{\mathfrak{D}_r^*} \tag{3}$$

is either a single point or a continuum. Suppose that $L'$ is another path in $D$ terminating at $z_0$. If, for every sufficiently small $r(>0)$, the last

[1] The order of connectivity of $D$ may be infinite.

[2] For any point set $M$, $\overline{M}$ always denotes its closure.

[3] Take as $D$ the unit circular disc with a radial slit and select a boundary point $z_0 (\neq 0)$ on the slit. If $w = f(z)$ is a function mapping $D$ conformally onto $|w| < 1$, then $C_D(f, z_0)$ consists of two points.

parts $L_r$ and $L'_r$ can be joined by a suitable path in $D_r$, then we say that $L$ and $L'$ are equivalent and define the same accessible boundary point of $D$ at $z_0$. It is easy to show that if $L$ and $L'$ are equivalent, then

$$C_D(f, z_0; L) = C_D(f, z_0; L') . \tag{4}$$

Evidently

$$C_D(f, z_0; L) \subset C_D(f, z_0) . \tag{5}$$

In the particular case where $D$ is a Jordan domain,

$$C_D(f, z_0) = C_D(f, z_0; L) . \tag{6}$$

(ii) The **Boundary Cluster Sets** $C_\Gamma(f, z_0)$ and $C_{\Gamma-E}(f, z_0)$. $\alpha \in C_\Gamma(f, z_0)$ [resp. $C_{\Gamma-E}(f, z_0)$] if there exists a sequence of points $\{\zeta_n\}$ of $\Gamma - z_0$ [resp. $\Gamma - z_0 - E$] such that

$$w_n \in C_D(f, \zeta_n) \text{ for each } n,$$
$$z_0 = \lim_{n\to\infty} \zeta_n \quad \text{and} \quad \alpha = \lim_{n\to\infty} w_n ;$$

i. e., if $M_r$ denotes the closure of the union $\bigcup_\zeta C_D(f, \zeta)$ for every $\zeta$ of the common part of $\Gamma - z_0$ [resp. $\Gamma - z_0 - E$] and $(c)$: $|z - z_0| < r$, then $\bigcap_{r>0} M_r$ is $C_\Gamma(f, z_0)$ [resp. $C_{\Gamma-E}(f, z_0)$]. Obviously $C_\Gamma(f, z_0)$ and $C_{\Gamma-E}(f, z_0)$ are closed;

$$C_{\Gamma-E}(f, z_0) \subset C_\Gamma(f, z_0) \subset C_D(f, z_0) ; \tag{7}$$

if $z_0 \in \Gamma - E$ or if $z_0 \in E - E'$, $E'$ denoting the derived set of $E$, then

$$C_{\Gamma-E}(f, z_0) = C\ \ (f, z_0) .$$

$C_\Gamma(f, z_0)$ is empty if and only if $z_0$ is an isolated boundary point; $C_{\Gamma-E}(f, z_0)$ is empty if and only if $z_0 \notin \overline{(\Gamma - E)}$.

(iii) The **Range of Values** $R_D(f, z_0)$. This is defined as the set of values $\alpha$ such that $z_n\ \ D$, $\lim_{n\to\infty} z_n = z_0$, $f(z_n) = \alpha$; i. e.,

$$R_D(f, z_0) = \bigcap_{r>0} \mathfrak{D}_r , \tag{8}$$

where $\mathfrak{D}_r$ is the value set of $w = f(z)$ in the common part of $D$ and $(c)$: $|z - z_0| < r$. Accordingly, $R_D(f, z_0)$ is a $G_\delta$ set.

(iv) The **Asymptotic Set** $A_D(f, z_0)$. Let $z_0$ be an accessible boundary point of $D$. A complex number $\alpha$ is called an *asymptotic value* of $w = f(z)$ at $z_0$ if $f(z) \to \alpha$ as $z \to z_0$ along a path in $D$ terminating at $z_0$. The asymptotic set $A_D(f, z_0)$ is defined as the set of asymptotic values of $f(z)$ at $z_0$. We define $A_D(f, z_0) = \emptyset$ when $z_0$ is an inaccessible boundary point, for the sake of convenience.

2. We shall state a relation between $C_D(f, z_0; L)$ and $C_{\Gamma-E}(f, z_0)$ which will be used later. If $z_0$ is an accessible boundary point of $D$ defined

by a path $L$ in $D$ terminating at $z_0$ and if $z_0$ is an accumulation point of $\Gamma - E$, then

$$C_D(f, z_0; L) \cap C_{\Gamma - E}(f, z_0) \neq \emptyset . \tag{9}$$

To prove this, let $\{\zeta_n\}$ be a sequence of points such that $\zeta_n \in \Gamma - E$ and $\zeta_n \to z_0$. Construct a simple closed curve $\gamma_n$ passing through $\zeta_n$ such that $\gamma_n$ surrounds $z_0$ and does not meet $E$. By a suitable choice of the sequence $\{\gamma_n\}$, we may assume that the diameter of $\gamma_n$ converges to zero. Let $z_n$ be the last point of intersection of $L$ with $\gamma_n$. Then, it is obvious that the component, containing $z_n$, of the intersection of $\gamma_n$ with $D$ is a cross-cut of $D$ whose end-points lie in $\Gamma - E$. From this fact follows that for every positive number $r$, $\bar{\mathfrak{D}}_r^*$ and $M_r$ (defined before) have a point in common and hence (9) holds.

## § 2. Some classical theorems

We recall some important classical theorems which will be made use of for the sequel.

**1.** Let $w = f(z)$ be a single-valued meromorphic function in a domain $D$: $0 < |z - z_0| < r$ which has an essential singularity at $z_0$. Then, it is well-known that

(i) $C_D(f, z_0)$ *is the whole w-plane* (Weierstrass' theorem);

(ii) *the complement* $\mathscr{C} R_D(f, z_0)$ *of* $R_D(f, z_0)$ *with respect to the w-plane contains at most two points* (Picard's theorem);

(iii) $\mathscr{C} R_D(f, z_0) \subset A_D(f, z_0)$ (Theorem of IVERSEN [1]).

**2. Iversen's theorems**[1]. Let $w = f(z)$ be a non-rational meromorphic function in $|z| < \infty$ and $z = \varphi(w)$ be its inverse analytic function. Let $c$: $|w - \alpha| = r$ be an arbitrary circle in the $w$-plane. Suppose that $e(w, w_0)$ is an arbitrary (regular or algebraic) element of $z = \varphi(w)$ with center $w_0$ lying in $(c)$: $|w - \alpha| < r$. IVERSEN [1] has proved that *it is possible to find a path* $\gamma_w$ *inside* $(c)$, *starting at* $w = w_0$ *and terminating at* $w = \alpha$, *such that there exists an analytic continuation of* $e(w, w_0)$ *of algebraic character*[2] *along* $\gamma_w$ *except perhaps the end-point* $w = \alpha$ *of* $\gamma_w$; we call this property Iversen's *property* or (I)-*property*[3]. It is easy to prove that (I)-property is equivalent to the property that given any element $e(w, w_0)$ of $z = \varphi(w)$, an arbitrary curve $\Lambda_w$, starting from $w = w_0$ and ending at $w = w_1$, and an arbitrary strip[4] $S$ containing $\Lambda_w$ completely in its

[1] NEVANLINNA [6], p. 291.

[2] Concerning notions of analytic continuation of algebraic character and (ordinary or essential) transcendental singularity, cf. COLLINGWOOD and CARTWRIGHT [1], pp. 99—103; NOSHIRO [4], pp. 43—73.

[3] Iversen's property of analytic functions has been systematically investigated by STOÏLOW [1, 9].

[4] $S$ denotes the union of all circular discs of constant radius and with center lying on $\Lambda_w$.

interior, we can find a path $L_w$, connecting $w_0$ and $w_1$, inside $S$, along which the analytic continuation of $e(w, w_0)$ is possible except perhaps at $w = w_1$.

Suppose that $w = f(z)$ has an asymptotic value $\alpha$ at $z = \infty$ along a curve $L_z$: $z = z(t)$, $0 \leqq t < 1$, $\lim_{t \to 1} z(t) = \infty$. Let $e_{z(t)}$ be the element of $z = \varphi(w)$ corresponding to $z(t)$. Then, the analytic continuation $\{e_{z(t)}, 0 \leqq \leqq t < 1\}$ of algebraic character along $L_w$: $w = w(t) = f(z(t))$, $0 \leqq t < 1$, $\lim_{t \to 1} w(t) = \alpha$ defines a transcendental (ordinary) singularity at $w = \alpha$. The converse is also true. If there exists an analytic continuation $\{e(w, w(t)), 0 \leqq t < 1\}$ along a path $L_w$: $w = w(t)$, $0 \leqq t < 1$, $\lim_{t \to 1} w(t) = \alpha$ which defines a transcendental singularity at $w = \alpha$, then $w = f(z)$ has an asymptotic value $\alpha$ at $z = \infty$ along the curve $L_z$: $z = z(t) = e(w(t), w(t))$, $0 \leqq t < 1$, $\lim_{t \to 1} z(t) = \infty$.

**3. Gross' star theorem**[1]. Let $w = f(z)$ be a non-rational meromorphic function and $z = \varphi(w)$ be its inverse. Let $e(w, w_0)$ be an arbitrary regular element of $z = \varphi(w)$. We continue analytically $e(w, w_0)$, using only regular elements, along every ray: $\arg(w - w_0) = \theta$ $(0 \leqq \theta < 2\pi)$ towards infinity. Then, there arise two cases whether the continuation defines a singularity $\omega_\theta$ in a finite distance or not; in the former case, we call the ray a *singular* ray. For each singular ray: $\arg(w - w_0) = \theta$, we exclude the segment between the singularity $\omega_\theta$ and $w = \infty$ from the $w$-plane. The remaining domain $\Delta_w$ is clearly a (simply connected) star domain in which the element $e(w, w_0)$ defines a (single-valued) regular branch of $z = \varphi(w)$. The star theorem of Gross [1] states that *the set of $\theta$ of singular rays*: $\arg(w - w_0) = \theta$ $(0 \leqq \theta < 2\pi)$ *is of measure zero*; i. e., $e(w, w_0)$ can be continued (with rational character) to infinity along almost all rays from the center $w_0$ (Gross' *property*).

It is easy to show that Iversen's theorem is a direct consequence of Gross' theorem; i. e., Iversen's property follows from Gross' property. Gross' property is more metrical and less topological than Iversen's property.

As an application of the Gross star theorem, we prove that *if $\alpha$ is an exceptional value in the sense of* Picard[2], *then $\alpha$ is an asymptotic value of $w = f(z)$ at $z = \infty$*. Without loss of generality, we may suppose that $\alpha = \infty$. Choose a point $w_0$ such that there exists an infinite number of elements $e_n(w, w_0)$ $(n = 1, 2, \ldots)$ of $z = \varphi(w)$ with center $w = w_0$. Then, by the Gross theorem, there exists at least one ray from $w = w_0$ along which every element $e_n(w, w_0)$ can be continued to infinity. Since there is only a

[1] Nevanlinna [6], p. 292.

[2] This means that $\alpha$ is taken by $w = f(z)$ only finite times in $|z| < \infty$.

finite number of elements of $z = \varphi(w)$ with center $w = \infty$, the continuation of some $e_n(w, w_0)$ defines a transcendental singularity at $w = \infty$[1].

**4.** We now enunciate some fundamental theorems on meromorphic functions in the unit circle.

**Theorem of FATOU**[2]. *Let $w = f(z)$ be regular and bounded in the unit circle $D$: $|z| < 1$. Then, $w = f(z)$ has an angular limit $f(e^{i\theta})$ at almost every point $z = e^{i\theta}$ of $\Gamma$: $|z| = 1$.*

**Theorem of F. and M. RIESZ**[3]. *Let $w = f(z)$ be regular and bounded in the unit circle $D$: $|z| < 1$. If the boundary function $f(e^{i\theta})$ is equal to $\alpha$ on a subset of positive measure of $\Gamma$, then $f(z) \equiv \alpha$.*

NEVANLINNA[4] has extended these theorems to the case of meromorphic functions of bounded type.

**Theorem of LINDELÖF-IVERSEN-GROSS**[5]. *Let $w = f(z)$ be a function, meromorphic in the unit circle $D$: $|z| < 1$, which omits three different values. If $w = f(z)$ has an asymptotic value $\alpha$ along a simple curve $L$ in $D$ terminating at $z_0 = e^{i\theta_0}$, then $f(z)$ has necessarily the angular limit $\alpha$ at $z_0 = e^{i\theta_0}$.*

**Theorem of KOEBE-GROSS**[6]. *Let $w = f(z)$ be a function, meromorphic in $|z| < 1$, which omits three different values in $|z| < 1$, and let there exist two sequences $\{z_n^{(1)}\}$ and $\{z_n^{(2)}\}$ such that $|z_n^{(1)}| < 1$, $\lim\limits_{n\to\infty} z_n^{(1)} = e^{i\theta_1}$; $|z_n^{(2)}| < 1$, $\lim\limits_{n\to\infty} z_n^{(2)} = e^{i\theta_2}$ where $\theta_1 \neq \theta_2$. If there is a sequence of continuous curves $\gamma_n$ joining $z_n^{(1)}$ to $z_n^{(2)}$ and contained in an annulus $1 - \varepsilon_n < |z| < 1$, where $\varepsilon_n > 0$, $\lim\limits_{n\to\infty} \varepsilon_n = 0$, such that on $\gamma_n$ we have $|f(z) - \alpha| < \eta_n$ where $\lim\limits_{n\to\infty} \eta_n = 0$, then $f(z) \equiv \alpha$.*

## II. Single-valued analytic functions in general domains

It belongs to one of the most important problems to study singularities, distribution of values, boundary-behaviours of analytic functions of a general domain of existence and their Riemann surfaces. In this chapter, we discuss mainly on single-valued analytic functions with a compact set of logarithmic capacity zero of essential singularities from the viewpoint of cluster sets[7].

---

[1] Modifying the argument slightly, we see that this result also holds in the case where $f(z)$ is a single-valued meromorphic function in $0 < |z - z_0| < r$ with an essential singularity at $z = z_0$; i. e. (iii) holds.

[2] FATOU [1].

[3] F. and M. RIESZ [1].

[4] NEVANLINNA [6], p. 208 and p. 209.

[5] PHRAGMÉN-LINDELÖF [1], IVERSEN [1], GROSS [1].

[6] KOEBE [1, 2]; GROSS [1], pp. 35—36.

[7] Nevanlinna's theory of meromorphic functions (in the parabolic case) has been extended independently by G. AF HÄLLSTRÖM [2] and TSUJI [3, 7] to the case of a compact set of logarithmic capacity zero of essential singularities.

## § 1. Compact set of capacity zero and Evans-Selberg's theorem

**1.** We recall some basic properties of a compact set of capacity zero[1]. Let $E$ be a bounded Borel set in the $z$-plane and $\mu$ be a non-negative completely additive set function defined for the Borel subsets of $E$. Then $\mu$ is called a positive mass-distribution on $E$. Let $\mu$ be a positive mass-distribution on $E$ with total mass unity. Then

$$U^\mu(z) = \int_E \log\left|\frac{1}{z-\zeta}\right| d\mu(\zeta) \tag{1}$$

is called a (logarithmic) potential of distribution $\mu$ on $E$. Writing

$$V_\mu(E) = \sup_z U^\mu(z)\,, \quad V = \inf_\mu V_\mu(E)\,, \tag{2}$$

we define the (logarithmic) capacity $C(E)$ of $E$ by

$$C(E) = e^{-V}.^{2} \tag{3}$$

Obviously $0 \leqq C(E) < \infty$; if $E_1 \subset E_2$, then $C(E_1) \leqq C(E_2)$; moreover, if there exists a sequence of bounded Borel sets $E_n$, such that $C(E_n) = 0$ for all $n$, and if $E = \bigcup_{n=1}^{\infty} E_n$ is bounded, then $C(E) = 0$.

**2.** Let us consider a domain $D$, containing $z = \infty$ in its interior, with boundary $\Gamma$. We suppose now that $E$ is a compact set complementary to $D$. Let $\{D_n\}$ be an exhaustion of $D$ such that each $D_n$ is bounded by a finite number of simple closed analytic curves $\Gamma_n$ and such that $\overline{D}_n \subset D_{n+1}$ $(n = 1, 2, \ldots)$. Denote by $g_n(z, \infty)$ Green's function of $D_n$ with pole at $z = \infty$. Since $\{g_n(z, \infty)\}$ is a monotone increasing sequence, the limit is either a finite function $g(z, \infty)$ in $D$ except for $z = \infty$ or a constant $\infty$. In the former case, $g(z, \infty)$ is called Green's function of $D$ with pole $z = \infty$ and in the latter we say that there exists no Green's function of $D$. It is well-known that there exists no Green's function of $D$ if and only if $C(E) = 0$[3].

**3.** Now, let $D_0$ be an arbitrary Jordan domain bounded by a simple closed analytic curve $\Gamma_0$ such that $\overline{D}_0 \subset D_1$. For simplicity, we put $G = D - \overline{D}_0$, $G_n = D_n - \overline{D}_0$. We denote by $\omega_n(z) = \omega(z, \Gamma_n, G_n)$ the harmonic measure with boundary values 0 on $\Gamma_0$ and 1 on $\Gamma_n$ respectively. Since $\{\omega_n(z)\}$ is monotonically decreasing, this sequence converges uniformly on any compact set in $G$ (Harnack's theorem); we denote the limiting function by

$$\omega(z) = \omega(z, \Gamma, G)\,.$$

[1] Throughout this book, "capacity" always means "logarithmic capacity". Logarithmic capacity, logarithmic potential and harmonic measure are discussed in details in Nevanlinna [6]. Concerning general potentials, cf. Frostmann [1], Kametani [4].

[2] In case $V = \infty$, we put $C(E) = 0$.

[3] Nevanlinna [6], p. 123.

Evidently, $\omega(z)$ is harmonic on $G \cup \Gamma_0$[1]; $\omega(z) = 0$ on $\Gamma_0$ and $0 \leqq \omega(z) < 1$ in $G$. By the minimum principle, if $\omega(z)$ vanishes at some point in $G$, then $\omega(z) \equiv 0$. If $\omega(z, \Gamma, G) \equiv 0$, then we say that $E$ is of absolute harmonic measure zero (NEVANLINNA)[2]. If $\Gamma$ contains a non-degenerate continuum, then $\omega(z, \Gamma, G) > 0$[3]. Accordingly, if $\Gamma$ is of absolute harmonic measure zero, then $\Gamma$ (and therefore $E$) is totally disconnected. Furthermore, $\Gamma$ is of absolute harmonic measure zero if and only if $C(E) = 0$[4].

Remark. Letting $z_i$ $(i = 1, 2, \ldots, n)$ vary on a compact set $E$, we denote by $V_n$ the maximum value of the quantity

$$V(z_1, z_2, \ldots, z_n) = \prod_{k<\lambda}^{1\cdots n} |z_k - z_\lambda| \,.$$

Then, $\sqrt[\binom{n}{2}]{V_n}$ is monotonically decreasing and converges to a limit $\tau(E)$ which is named by FEKETE [1] the transfinite diameter of $E$. It is known that $C(E) = \tau(E)$[5].

**4.** We add a remark on metrical properties of a compact set of capacity zero[6]. Let $E$ be a compact set. If for any positive number $\varepsilon$, we can cover $E$ by a sequence of circular discs $K_n$ of radius $r_n$ such that $\Sigma r_n < \varepsilon$, then we say that $E$ is of linear measure zero. Similarly, we define $E$ to be of logarithmic measure zero, by replacing $\Sigma r_n < \varepsilon$ by $\Sigma (\log^+ 1/r_n)^{-1} < \varepsilon$. It is known that if $E$ is of logarithmic measure zero, then $C(E) = 0$; if $C(E) = 0$, then $E$ is of linear measure zero; their converses are not true.

**5. Evans-Selberg's theorem.** G. C. EVANS [1] and H. SELBERG [1] have proved independently the following

**Theorem 1.** *Let $E$ be a compact set of capacity zero. Then there exists a positive mass-distribution $\mu$ on $E$ with total mass unity, such that its potential*

$$u(z) = \int_E \log \left| \frac{1}{z - \zeta} \right| d\mu(\zeta) \tag{4}$$

*is positively infinite at every point of $E$ and at no other points.*

*Proof*[7]. Given $n$ points $a_1, a_2, \ldots, a_n$ on $E$, we form a polynomial $P(z) = (z - a_1)(z - a_2) \ldots (z - a_n)$. Denote by $\overline{M}_n$ the maximum modulus

---

[1] Since $\omega_n(z) = 0$ on $\Gamma_0$, it follows, by Schwarz's principle of reflection, that $\omega(z)$ is also harmonic on $\Gamma_0$.

[2] The distinction whether $\omega(z, \Gamma, G)$ identically vanishes or not is independent of the choice of an exhaustion $\{D_n\}$ $(n = 0, 1, 2, \ldots)$ of $D$. Cf. NEVANLINNA [6], p. 119.

[3] NEVANLINNA [6], p. 120.

[4] NEVANLINNA [6], p. 126.

[5] NEVANLINNA [6], p. 135.

[6] Cf. NEVANLINNA [6], pp. 148—163; also KAMETANI [4].

[7] NOSHIRO [6], G. AF HÄLLSTRÖM [2]. This proof is essentially the same as that of EVANS [1], although Evans' original theorem is stated in the case of 3-dimensions.

of $P(z)$, letting $z$ vary on $E$, i. e., $\overline{M}_n = \max_{z \in E} |P(z)|$, and by $M_n$ the greatest lower bound of $\overline{M}_n$, letting $n$ points $a_1, a_2, \ldots, a_n$ vary on $E$, i. e., $M_n = \inf \overline{M}_n$. Then, it is easily shown that $M_n$ is the minimum of $\overline{M}_n$; in other words, by a suitable choice of $a_1^0, a_2^0, \ldots, a_n^0$ on $E$, there exists a polynomial

$$T_n(z) = (z - a_1^0)(z - a_2^0) \ldots (z - a_n^0)$$

with maximum modulus $M_n$. Remembering the definition of the transfinite diameter $\tau(E)$ of $E$ and the relation $\tau(E) = C(E)$, we denote by $V_n$ the maximum of

$$V(z_1, z_2, \ldots, z_n) = \prod_{k<\lambda}^{1\cdots n} |z_k - z_\lambda| ,$$

letting $z_i$ $(i = 1, 2, \ldots, n)$ vary on $E$. Let the maximum $V_{n+1}$ be attained by $n+1$ points $b_1, b_2, \ldots, b_{n+1}$ on $E$. From

$$\begin{aligned} V_{n+1} &= V(b_1, b_2, \ldots, b_{n+1}) \\ &= |(b_1 - b_2)(b_1 - b_3) \ldots (b_1 - b_{n+1})| \cdot V(b_2, b_3, \ldots, b_{n+1}) \end{aligned}$$

follows

$$|(b_1 - b_2)(b_1 - b_3) \ldots (b_1 - b_{n+1})| \geqq M_n , \tag{5}$$

for otherwise there would exist a point $b_1' \in E$ such that $V(b_1, b_2, \ldots, b_{n+1}) < V(b_1', b_2, \ldots, b_{n+1})$. By a cyclic change of suffices of $b$ in (5), we have

$$V_{n+1} \geqq M_n^{\frac{n+1}{2}} \quad \text{and} \quad \sqrt[\binom{n+1}{2}]{V_{n+1}} \geqq \sqrt[n]{M_n},$$

whence follows

$$\lim_{n\to\infty} \sqrt[\binom{n+1}{2}]{V_{n+1}} = \lim_{n\to\infty} \sqrt[n]{M_n} = 0,$$

as $\tau(E) = C(E) = 0$.

Consider now the function

$$\begin{aligned} u_n(z) &= -\log \sqrt[n]{|T_n(z)|} \\ &= \frac{1}{n}\left(\log\left|\frac{1}{z - a_1^0}\right| + \log\left|\frac{1}{z - a_2^0}\right| + \cdots + \log\left|\frac{1}{z - a_n^0}\right|\right) ; \end{aligned}$$

$u_n(z)$ is clearly a potential defined by a certain distribution of equal point masses on $E$ with total mass unity and for every point $z$ on $E$, $u_n(z) \geqq m_n$ where $m_n = -\log \sqrt[n]{M_n}$. Since $m_n \to \infty$, we can find a sequence of integers $\{n_j\}$ such that $m_{n_j} \geqq 2^j$ $(j = 1, 2, \ldots)$. Put $U_j(z) = 2^{-j} u_{n_j}(z)$ $(j = 1, 2, \ldots)$. Then, $U_j(z)$ is a potential of distribution of equal point masses on $E$ with total mass $2^{-j}$ and evidently $U_j(z) \geqq 1$ on $E$. Consider finally the function

$$u(z) = \sum_{j=1}^{\infty} U_j(z) = \lim_{\nu\to\infty} \sum_{j=1}^{\nu} U_j(z) .$$

Then, $u(z)$ is a required potential. In fact, it is a potential of positive mass-distribution on $E$ with total mass unity and hence of the form (4). At every point $z$ of $E$, $u(z) = +\infty$ as $u(z) \geqq \sum_{j=1}^{\nu} U_j(z) \geqq \nu$ for all $\nu$. If $z \in \mathscr{C}E$ and if $z$ has a distance $\varrho$ from $E$, then clearly $u(z) \leqq \log 1/\varrho$.

Remark. For convenience, we shall call the potential $u(z)$, in Theorem 1, an Evans-Selberg's *potential*. For a given compact set of capacity zero, Evans-Selberg's potential is not unique (G. AF HÄLLSTRÖM [3]).

**6.** For the sequel, it will be convenient to state some properties of Evans-Selberg's potential $u(z)$. Clearly $u(z)$ is harmonic outside $E$ except for $z = \infty$ and its boundary value at every point of $E$ is $+\infty$. In the neighborhood of $z = \infty$, $u(z)$ is of the form

$$u(z) = -\log|z| - \omega(z)\,, \tag{6}$$

where $\omega(z) = \int_E \log|1 - \zeta/z|\,d\mu(\zeta)$ is harmonic at $z = \infty$. Let $v(z)$ be its conjugate harmonic function and put

$$w(z) = u(z) + i\,v(z)\,. \tag{7}$$

Then the function $w(z)$ is many-valued and regular outside $E$ except for $z = \infty$, the infinity being a logarithmic singularity. However the derivative $w'(z) = u_x(z) - i\,u_y(z)$ is obviously single-valued and regular throughout the domain $\mathscr{C}E$, $z = \infty$ being a simple zero-point of $w'(z)$, and has a singularity at every point of $E$. Consequently, the many-valuedness of $w(z)$ arises only in its imaginary part by some additive constants. It is easy to show that the level curve $\Gamma_\lambda$: $u(z) = \lambda$ $(-\infty < \lambda < \infty)$ consists of a finite number of simple closed curves surrounding $E$, by the minimum principle of harmonic function, and that the function $\lambda - u(z)$ is no other than Green's function $g(z, \infty)$ in the exterior of $\Gamma_\lambda$. Thus we see that if there are $p$ closed curves of $\Gamma_\lambda$, then $w'(z)$ has $p - 1$ finite zero-points in the exterior of $\Gamma_\lambda$ and moreover that

$$\int_{\Gamma_\lambda} dv(z) = \int_{\Gamma_\lambda} \frac{\partial u}{\partial n}\,ds = 2\pi\,, \tag{8}$$

where $ds$ denotes the arc length and $n$ the inner normal (see HÄLLSTRÖM [2], pp. 14—17).

Remark. Recently extensions of the Evans-Selberg theorem and related theorems have been obtained by RUDIN [1], UGAERI [1], HONG [1] and INOUE [1].

## § 2. Meromorphic functions with a compact set of essential singularities of capacity zero

**1.** At the beginning, we prove

**Theorem 1.** *Let $E$ be a compact set of capacity zero and $D$ be a domain containing $E$ in its interior. Suppose that $w = f(z)$ is a single-valued mero-*

*morphic function in $D - E$ and has a transcendental singularity at every point $z_0$ of $E$. Then, the cluster set $C_{D-E}(f, z_0)$ of $f(z)$ at $z = z_0$ is the whole $w$-plane* (NEVANLINNA [6]).

*Proof.* Obviously we have only to prove that $f(z) = u(z) + iv(z)$ is not bounded in any neighborhood of every point $z_0$ of $E$. Otherwise, $f(z)$ would be bounded in the intersection of $D - E$ with a circular disc $(c)$: $|z - z_0| < r$. Describe a simple closed curve $\Gamma$, surrounding $z_0$, in $(D - E) \cap (c)$[1] and denote by $\Delta$ the remaining domain obtained by excluding $E$ from the interior of $\Gamma$. Let $\bar{u}(z)$ be the harmonic function in the interior of $\Gamma$, such that $\bar{u}(z) = u(z)$ on $\Gamma$, and $u^*(z)$ be the Evans-Selberg potential which may be supposed to be positive in $(c)$. Consider $U(z) = u(z) - \bar{u}(z) - \varepsilon u^*(z)$ in $\Delta$ for any positive number $\varepsilon$. Then, clearly $U(z) \leqq 0$ in $\Delta$; hence $u(z) \leqq \bar{u}(z)$ in $\Delta$. Similarly $\bar{u}(z) \leqq u(z)$ in $\Delta$. Thus $u(z) = \bar{u}(z)$ in $\Delta$. Accordingly $z_0$ is a removable singularity for $f(z)$; this is a contradiction.

Remark. In the proof, we have used only the fact that if a harmonic function is bounded in a neighborhood of a compact set of capacity zero, this set is removable for the harmonic function. Obviously Theorem 1 remains valid if $E$ is a Painlevé *null-set*[2], i. e. if $E$ consists of *A B removable points*.

**Theorem 2.** *Let $E$ be a compact set of capacity zero contained in a domain $D$. Suppose that $w = f(z)$ is a single-valued meromorphic function in $D - E$ which has an essential singularity at every point $z_0$ of $E$. Then, $w = f(z)$ assumes every value infinitely often in any neighborhood of $z_0$ with a possible exceptional set of values of capacity zero; i. e. $\mathscr{C} R_{D-E}(f, z_0)$ is at most of capacity zero* (G. AF HÄLLSTRÖM [2], KAMETANI [2]).

*Proof.* By Theorem 1, $R_{D-E}(f, z_0)$ is everywhere dense in the $w$-plane. Without loss of generality, we may suppose that $w = \infty$ belongs to $R_{D-E}(f, z_0)$. Let $r$ be any positive number and $(c)$ be a circular disc $|z - z_0| < r$. Describe a simple closed curve $\Gamma$, surrounding $z_0$, in $(D - E) \cap (c)$. Denote by $\Delta_r$ the domain $(\Gamma) - E$, where $(\Gamma)$ is the interior of $\Gamma$, and by $\mathfrak{D}_r$ the value set of $f(z)$ in $\Delta_r$. It is easily shown that the compact set $\mathscr{C}\mathfrak{D}_r$, which is complementary to $\mathfrak{D}_r$ with respect to the $w$-plane, does not contain any non-degenerate continuum; i. e. $\mathscr{C}\mathfrak{D}_r$ is totally disconnected. We show that $\mathscr{C}\mathfrak{D}_r$ is of capacity zero. Otherwise, there would exist a non-constant bounded harmonic function $U(w)$ in $\mathfrak{D}_r$. We

[1] As $E$ is of capacity zero, $E$ is of linear measure zero. Consequently we can adopt as $\Gamma$ a circumference $|z - z_0| = \varrho$ for almost every positive number $\varrho < r$. But our selection of $\Gamma$ depends upon only the property that $E$ is totally disconnected.

[2] If there exists no non-constant single-valued bounded analytic function in the exterior of a totally disconnected compact set $E$, then $E$ is called a Painlevé null-set or said to consist of $A$ $B$ removable points. It is easily proved that if $E$ is of linear measure zero, then $E$ is a Painlevé null-set (AHLFORS [3], AHLFORS-BEURLING [1]). For related theorems, cf. A. S. BESICOVITCH [1], CARTWRIGHT [3].

suppose that $U(w_1) \neq U(w_2)$ for two points $w_1, w_2$ in $\mathfrak{D}_r$.[1] Consider the composed function $H(z) = U(f(z))$ in $\Delta_r$. Since $H(z)$ is bounded and harmonic in $\Delta_r$, each point of $E$ situated in $(\Gamma)$ is removable for $H(z)$ (cf. the preceding remark). On the other hand, there exist two sequences of points $\{z_\nu^{(1)}\}$, $\{z_\nu^{(2)}\}$ in $\Delta_r$ such that $z_\nu^{(1)} \to z_0$, $z_\nu^{(2)} \to z_0$ and $f(z_\nu^{(1)}) \to w_1$, $f(z_\nu^{(2)}) \to w_2$. Then, it holds that $H(z_\nu^{(1)}) = U(f(z_\nu^{(1)})) \to U(w_1)$, $H(z_\nu^{(2)}) = U(f(z_\nu^{(2)})) \to U(w_2)$, contradicting the remark of Theorem 1. Thus, $\mathscr{C}\mathfrak{D}_r$ is of capacity zero. Now, we consider a sequence of circles $K_n$: $|z - z_0| = 1/n$ $(n = 1, 2, \ldots)$, its interior being denoted by $(K_n)$. Denote by $\{e_n\}$ the set of values omitted by $w = f(z)$ in $(D - E) \cap (K_n)$; then, $\{e_n\}$ is a monotone increasing sequence of compact sets. Our assertion follows from the inclusion $\mathscr{C} R_{D-E}(f, z_0) \subset \bigcup_{n=1}^{\infty} e_n$, since $e_n \subset \mathscr{C}\mathfrak{D}_r$ provided that $r < 1/n$.

Remark. Theorem 2 can be extended to the case when $E$ is a Painlevé null-set. In this case, the assertion states that either $\mathscr{C} R_{D-E}(f, z_0)$ is empty or any compact subset of $\mathscr{C} R_{D-E}(f, z_0)$ is a Painlevé null-set. To prove this, we note the fact that if a compact set $e$ is contained in the union of compact Painlevé null-sets $\bigcup_{n=1}^{\infty} e_n$, then $e$ is also a Painlevé null-set[2]; and then we have only to replace two words "of capacity zero" and "harmonic" by "a Painlevé null-set" and "analytic" in the preceding proof[3].

2. Let $E$ be a compact set of capacity zero and $D$ be its complement. Let $u(z)$ be an Evans-Selberg's potential of positive mass-distribution $\mu$ of total mass unity on $E$; i. e. $u(z) = \int_E \log |1/(z - a)| \, d\mu(a)$, and $v(z)$ be its conjugate. For convenience, we call the function

$$\zeta = \chi(z) = e^{u(z) + iv(z)} = \varrho(z)\, e^{iv(z)} \tag{1}$$

an Evans-Selberg's *function*. As already stated, the level curve $\Gamma_\lambda$: $\varrho(z) = \text{const.} = \lambda$ $(0 < \lambda < \infty)$ consists of a finite number of simple closed

[1] The existence of $U(w)$ is shown as follows. If $\mathscr{C}\mathfrak{D}_r$ is of positive capacity, there are two disjoint compact subsets $e_1$ and $e_2$ of $\mathscr{C}\mathfrak{D}_r$ of positive capacity. Let $g(w, \infty, G_i)$ be Green's function with pole at infinity of $G_i$ which is the complement of $e_i$ $(i = 1, 2)$. Then $U(w) = g(w, \infty, G_1) - g(w, \infty, G_2)$ is non-constant bounded and harmonic in $G_1 \cap G_2$ which contains $\mathfrak{D}_r$.

[2] This fact was pointed out by Kametani. Evidently $e$ is totally disconnected. If $e$ were not a Painlevé null-set, there would exist a non-constant single-valued bounded analytic function $\varphi(w)$ in the complement of $e$. Let $P$ be the perfect set of non-removable singularities of $\varphi(w)$. Then, there would exist a point $w_0 \in P$ such that the intersection of $P$ with some circular disc with center at $w_0$ is included in some $e_n$; this is clearly contradictory.

[3] Cf. also Cartwright [3], Ahlfors-Beurling [1].

curves surrounding $E$. Furthermore $\zeta = \chi(z)$ has the following important property

$$\int_{\Gamma_\lambda} dv(z) = 2\pi . \qquad (2)$$

**Theorem 3.** *Suppose that $w = f(z)$ is a single-valued meromorphic function in $D$ with a compact set $E$ of capacity zero of essential singularities. Let $\Phi$ be the Riemannian image of $D$ by $w = f(z)$. Then $\Phi$ has Gross' property* (Tsuji [3]).

*Proof.* Suppose that $P_0$ is a regular point of $\Phi$ above $w = w_0$. Consider the star domain $H$ in the sense of Gross formed by the union of segments from $P_0$ to singular points (algebraic branch-points or accessible boundary points) of $\Phi$ along all rays: $\arg(w - w_0) = \psi$. We show that the linear measure of the set of arguments $\psi$ of singular rays is equal to zero[1].

Denote by $H_R$ the part of $H$ above a circular disc $|w - w_0| < R$ $(<\infty)$ and by $\Delta_R$ the counter-image of $H_R$ by $w = f(z)$. Then $\Delta_R$ is a simply connected domain included in $D$. By the monodromy theorem, $\zeta = \chi(z)$ defines a single-valued regular function in $\Delta_R$. Now we consider the Riemannian image $\tilde{\Delta}_R$ of $\Delta_R$ by $\zeta = \chi(z)$ and the composed function $w = W(\zeta) \equiv f(\chi^{-1}(\zeta))$ defined on the covering surface $\tilde{\Delta}_R$. Let $\tilde{\Theta}_\lambda$ be the set of cross-cuts of $\tilde{\Delta}_R$ lying above the circle $|\zeta| = \lambda$. We denote by $\lambda\,\theta(\lambda)$ the total length of $\tilde{\Theta}_\lambda$ and by $L(\lambda)$ that of the image of $\tilde{\Theta}_\lambda$ by $w = W(\zeta)$. Then, applying a well-known method in proving the classical Gross star theorem, we get

$$\int_{\lambda_0}^{\lambda} \frac{[L(\lambda)]^2}{\lambda\,\theta(\lambda)}\, d\lambda \leqq \int_{\lambda_0}^{\lambda} \int_{\tilde{\Theta}_\lambda} |W'(\zeta)|^2\, \lambda\, d\lambda\, d\theta \leqq \pi R^2 , \qquad (3)$$

$(0 < \lambda_0 < \lambda)$. Since $\theta(\lambda) \leqq 2\pi$ by (2), we have $\varliminf_{\lambda\to\infty} L(\lambda) = 0$; accordingly, it follows that the set of $\arg(w - w_0) = \psi$ of singular rays is of measure zero[2].

It seems very important to remark that we can apply Ahlfors' theory of covering surfaces[3] in the present case, with an aid of Evans-Selberg's function. Now we denote by $\Delta_\lambda$ the exterior of the level curve $\Gamma_\lambda$: $\varrho(z) = \lambda$ $(0 < \lambda < \infty)$ and by $\Phi_\lambda$ the Riemannian image of $\Delta_\lambda$ by $w = f(z)$ in Theorem 3. Denote now by $A(\lambda)$ the spherical area of $\Phi_\lambda$ and by $L(\lambda)$ the spherical length of the image of $\Gamma_\lambda$. Then we have

$$\text{(i)}\ \lim_{\lambda\to\infty} A(\lambda) = \infty \quad \text{and} \quad \text{(ii)}\ \varliminf_{\lambda\to\infty} L(\lambda)/A(\lambda) = 0.$$

---

[1] We understand by a singular ray a half straight line starting from $P_0$ and meeting a singular point of $\Phi$ in a finite distance.

[2] Cf. Nevanlinna [6], p. 292.

[3] Ahlfors [2].

Evidently (i) follows from Theorem 2. To prove (ii), we rewrite

$$A(\lambda) = \iint\limits_{\Delta_\lambda} \frac{|f'(z)|^2}{(1+|f(z)|^2)^2}\, d\sigma \quad (d\sigma\text{: the area element})\,,$$

$$L(\lambda) = \int\limits_{\Gamma_\lambda} \frac{|f'(z)|}{1+|f(z)|^2}\, |dz|\,,$$

by using a suitable branch[1] of Evans-Selberg's function: $\zeta = \chi(z) = e^{u(z)+iv(z)}$, $0 \leqq v(z) < 2\pi$. If we set $w = W(\zeta) \equiv f(\chi^{-1}(\zeta))$, then

$$A(\lambda) - A(\lambda_0) = \int\limits_{\lambda_0}^{\lambda} \int\limits_{\tilde{\Gamma}_\lambda} \frac{|W'(\zeta)|^2}{(1+|W(\zeta)|^2)^2}\, \lambda\, d\lambda\, d\theta \quad (\zeta = \lambda\, e^{i\theta})\,,$$

where $\tilde{\Gamma}_\lambda$ denotes the image, on $|\zeta| = \lambda$, of $\Gamma_\lambda$ by the branch of $\zeta = \chi(z)$, and

$$L(\lambda) = \int\limits_{\tilde{\Gamma}_\lambda} \frac{|W'(\zeta)|}{1+|W(\zeta)|^2}\, \lambda\, d\theta\,.$$

Making use of Schwarz's inequality,

$$\frac{d\lambda}{\lambda\,\theta(\lambda)} \leqq \frac{dA(\lambda)}{[L(\lambda)]^2}\,, \tag{4}$$

where $\lambda\,\theta(\lambda)$ denotes the length of $\tilde{\Gamma}_\lambda$; obviously $\lambda\,\theta(\lambda) = 2\pi\lambda$ by (2). From (4) follows (ii). In other words, we get

**Theorem 4.** *The covering surface $\Phi$ is regularly exhaustible in the sense of* Ahlfors (Noshiro [6]).

Consequently, using an exhaustion $\{\Phi_\lambda\}$ of $\Phi$, we obtain

**Theorem 5.** *Let $D_1, D_2, \ldots, D_q$ $(q \geqq 3)$ be $q$ closed disjoint circular discs on the Riemann $w$-sphere. Define the defect $\delta(D_j)$, the ramification index $\vartheta(D_j)$ and a quantity $\xi$ by*

$$\delta(D_j) = \varliminf_{\lambda\to\infty} \left[1 - \frac{n(\lambda, D_j)}{S(\lambda)}\right], \quad \vartheta(D_j) = \varliminf_{\lambda\to\infty} \frac{n_1(\lambda, D)}{S(\lambda)}\,, \quad \xi = \varlimsup_{\lambda\to\infty} \frac{\varrho^+(\Delta_\lambda)}{S(\lambda)}\,,$$

*where $n(\lambda, D_j)$ denotes the number of sheets of all islands of $\Phi_\lambda$ above $D_j$, $n_1(\lambda, D_j)$ the total number of orders of the branch-points of the islands*

[1] Although $\zeta = \chi(z)$ is many-valued, we use a suitable single-valued branch of $\zeta = \chi(z)$. If $R$ is sufficiently large, any branch of $\zeta = \chi(z)$ maps $|z| < R$ conformally upon a Jordan domain containing $\zeta = \chi(\infty) = 0$ in a one-to-one manner. Starting from the element $e(\zeta, 0)$ at the origin $\zeta = 0$ of the inverse $z = \chi^{-1}(\zeta)$, we define a star-domain $H_\zeta$ in the sense of Gross. It is easily proved that the boundary of $H_\zeta$ consists of at most a countable number of infinite radial segments which accumulate only to $\zeta = \infty$, and that the element $e(\zeta, 0)$ defines a univalent meromorphic function $z = \varphi(\zeta)$ in $H_\zeta$. Clearly $z = \varphi(\zeta)$ maps $H_\zeta$ conformally onto a subdomain $D'$ of $D$ which is obtained from $D$ by making suitable slits along $v$-lines: $v(z) = $ constant. Then the branch of $\zeta = \chi(z)$, which is the inverse of $z = \varphi(\zeta)$, is a univalent regular function in $D'$; we denote this branch by $\zeta = \chi(z) = \exp(u(z) + iv(z))$, $(0 \leqq v(z) < 2\pi)$.

*of $\Phi_\lambda$ above $D_j$, $S(\lambda)$ the average number of sheets of $\Phi_\lambda$ with respect to the $w$-sphere, $\varrho(\Delta_\lambda)$ the Euler characteristic of $\Delta_\lambda$ and* $\varrho^+ = \max(0, \varrho)$. *Then, it holds*

$$\sum_{j=1}^{q} \delta(D_j) + \sum_{j=1}^{q} \vartheta(D_j) \leqq 2 + \xi\text{[1]}.$$

## § 3. Extension of Iversen's theorem on asymptotic values

**1.** First we state

**Lemma.** *Let $w = f(z)$ be single-valued and regular in a bounded domain $D$ and let $E$ be a closed set of capacity zero included in the boundary $\Gamma$ of $D$. If $\overline{\lim\limits_{z\to\zeta}}|f(z)| \leqq M$ for every point $\zeta$ of $\Gamma - E$ and if $f(z)$ is bounded in a neighborhood of every point of $E$, then $|f(z)| \leqq M$ for all points in $D$* (Extension of the maximum principle).

*Proof.* We suppose, contrary to the assertion, that there exists a point $z_0$ in $D$ such that $|f(z_0)| > M$. Let $\Phi$ be the Riemannian image of $D$ by $w = f(z)$ and denote by $P_0$ the point on $\Phi$ which corresponds to $z_0$. Consider the star domain in the sense of Gross formed by the union of segments from $P_0$ with projection $w_0 = f(z_0)$ to singular points along all rays: $\arg(w - w_0) = \psi$ on $\Phi$, whose projections lie in the half-plane $\Re[e^{-i \arg w_0} \cdot (w - w_0)] > 0$. With a slight modification of the proof for Theorem 3, § 2, we can prove that the set of $\arg(w - w_0) = \psi$ $(0 \leqq \psi < < 2\pi)$ of singular rays is of linear measure zero. Accordingly there exists at least one asymptotic path $\Lambda$ inside $D$ terminating at a point $\zeta$ in $E$, along which $w = f(z)$ converges to $\infty$ as $z$ tends to $\zeta$. But this is contradictory, since $f(z)$ is bounded in a neighborhood of $\zeta$.

Remark. This lemma is an immediate consequence of Nevanlinna's theorem[2]. However, it is easily proved that this lemma also holds in the case of pseudo-analytic functions of bounded eccentricity, by modifying slightly our proof[3].

**2.** We give an extension of Iversen's theorem which will be useful in the theory of cluster sets.

**Theorem 1.** *Let $D$ be an arbitrary domain, $\Gamma$ its boundary, $E$ a compact set of capacity zero on $\Gamma$ and $z_0$ a point of $E$. Suppose that $w = f(z)$ is single-valued and meromorphic in $D$, and $C_D(f, z_0) - C_{\Gamma - E}(f, z_0)$ is not empty. If $\alpha \in C_D(f, z_0) - C_{\Gamma - E}(f, z_0)$ is an exceptional value of $w = f(z)$ in a neighborhood of $z_0$, then either $\alpha$ is an asymptotic value of $f(z)$ at $z_0$ or there exists a sequence $\zeta_n \in E$ $(n = 1, 2, \ldots)$ converging to $z_0$ such that $\alpha$ is an asymptotic value of $f(z)$ at each $\zeta_n$* (Noshiro [2]).

---

[1] Ahlfors [2], Kunugui [3], Tumura [1, 2], Tsuji [1], Noshiro [6].

[2] Nevanlinna [6], p. 140.

[3] Noshiro [8].

*Proof.* It is sufficient to consider the case that $z_0$ is an accumulation point of $\Gamma - E$. By assumption, there exists an arbitrarily small positive number $r$ such that

$$K \cap E = \theta \quad \text{and} \quad f(z) \neq \alpha \quad \text{in} \quad \overline{(K)} \cap D$$

where $K$: $|z - z_0| = r$ and $\overline{(K)}$: $|z - z_0| \leqq r$. Moreover, we may suppose that $\alpha$ does not belong to $M_r$ which is the closure of the union $\bigcup_\zeta C_D(f, \zeta)$ for $\zeta$ belonging to $(\Gamma - E) \cap \overline{(K)}$. Let $\varrho_1$ be the distance of $\alpha$ from $M_r$. Next, we suppose that $|f(z) - \alpha| \geqq \varrho_2 > 0$ on $K \cap D$. Let $\varrho$ be a positive number less than $\min(\varrho_1, \varrho_2)$. Since $\alpha$ is a cluster value of $w = f(z)$ at $z_0$, there exists a sequence of points $z_n$ $(n = 1, 2, \ldots)$ inside $(K) \cap D$ converging to $z_0$ such that $w_n = f(z_n) \to \alpha$ where $(K)$ denotes the interior of $K$. Now, we consider the counter-image $D_0$ of $(c)$: $|w - \alpha| < \varrho$ inside $(K) \cap D$. Choose a point $w_n \in (c)$ and denote by $\Delta_0$ the connected component of $D_0$ containing $z_n$. Then the boundary of the domain $\Delta_0$ consists of a closed subset $E_0$ of $E$ and at most a countable number of analytic curves $\gamma_0$ [boundary relative to the open set $(K) \cap D$]. Obviously $f(z) - \alpha$ is regular in the closed domain $\overline{\Delta}_0$, except for the set $E_0$, and satisfies $0 < |f(z) - \alpha| < \varrho$ in $\Delta_0$ and $|f(z) - \alpha| = \varrho$ on the relative boundary $\gamma_0$ of $\Delta_0$. Applying the method in proving the preceding Lemma to the function $F(z) = (f(z) - \alpha)^{-1}$, we see that $F(z)$ has the asymptotic value $\infty$ at some point $z'$ belonging to $E_0$. Thus it is concluded that $w = f(z)$ has the asymptotic value $\alpha$ at $z_0'$.

As an immediate corollary, we get

**Theorem 2.** *Let $E$ be a compact set of capacity zero and $D$ be a domain containing $E$ in its interior. Suppose that $w = f(z)$ is a single-valued meromorphic function in $D - E$ and has an essential singularity at every point $z_0$ of $E$. If $\alpha$ is an exceptional value of $w = f(z)$ in a neighborhood of $z_0$, then either $\alpha$ is an asymptotic value of $w = f(z)$ at $z_0$ or there exists a sequence $\zeta_n \in E$ $(n = 1, 2, \ldots)$ converging to $z_0$ such that $\alpha$ is an asymptotic value at each $\zeta_n$* (CARTWRIGHT [2])[1].

Remark. It is open whether Theorem 2 holds in the case that $E$ is a Painlevé null-set.

## § 4. Extension of Iversen-Gross-Seidel-Beurling's theorem

**1.** We start with the following

**Theorem 1.** *Let $D$ be a bounded open set, $\Gamma$ its boundary, $E$ a compact set of capacity zero on $\Gamma$ and $z_0$ a point of $E$. Suppose that $z_0$ is a regular point for the Dirichlet problem*[2]. *If $u(z)$ is bounded from above and subharmonic in that part of $D$ contained in a neighborhood $U(z_0)$ of $z_0$, then it holds that*

$$\overline{\lim_{z \to z_0}}\, u(z) \leqq \overline{\lim_{\substack{\zeta \to z_0 \\ \zeta \in \Gamma - E}}} \left( \overline{\lim_{z \to \zeta}}\, u(z) \right). \tag{1}$$

[1] In the present case, $C_{D-E}(f, z_0)$ is the whole $w$-plane and $C_{\Gamma - E}(f, z_0)$ is empty.

[2] It follows that $z_0$ is an accumulation point of $\Gamma - E$.

*Proof.* Since $z_0$ is a regular point for the Dirichlet problem, Wiener's function $H(z)$ for the subharmonic function $|z - z_0|$ satisfies $H(z) = H_{|z-z_0|}(z) \geqq |z - z_0|$ in $D$ and $\lim_{z \to z_0} H(z) = 0$ (BRELOT [2]). Denote by $m$ the value of the right-hand side of (1). We may assume that $u(z) < M$ in $U(z_0) \cap D$ where $M$ is a fixed positive number greater than $m + 1$. For a given positive number $\varepsilon$ $(0 < \varepsilon < 1)$, we can find a neighborhood $V(z_0)$ of $z_0$ contained in $U(z_0)$ such that

$$\overline{\lim_{z \to \zeta}}\, u(z) < m + \varepsilon < M$$

at every point $\zeta \in (\Gamma - E) \cap V(z_0)$. Next, we describe a circle $K$: $|z - z_0| = r$ in $V(z_0)$ and denote by $D_r$ the intersection of $D$ with $(K)$: $|z - z_0| < r$. We consider a subharmonic function

$$\tilde{u}(z) = u(z) - \eta\, u^*(z) - \frac{M - (m + \varepsilon)}{r} H(z)$$

in $D_r$ where $\eta$ is an arbitrary positive number and $u^*(z)$ is an Evans-Selberg's potential associated with $E$ (cf. § 1). We may assume that $u^*(z)$ is positive in $D_r$. Since $\overline{\lim_{z \to \zeta}}\, \tilde{u}(z) \leqq m + \varepsilon$ at every boundary point $\zeta$ of $D_r$, we have $\tilde{u}(z) \leqq m + \varepsilon$ in $D_r$, i. e.,

$$u(z) \leqq \eta\, u^*(z) + \frac{M - (m + \varepsilon)}{r} H(z) + m + \varepsilon\,.$$

Keeping $z$ fixed and letting $\eta \to 0$, we have

$$u(z) \leqq \frac{M - (m + \varepsilon)}{r} H(z) + m + \varepsilon\,.$$

Next, if $z \to z_0$, then $\overline{\lim_{z \to z_0}}\, u(z) \leqq m + \varepsilon$. Thus our assertion is proved since $\varepsilon$ is an arbitrary positive number.

Now, we state an important theorem on cluster sets which is an extension of Iversen-Gross-Seidel-Beurling's theorem.

**Theorem 2.** (Iversen-Tsuji's theorem). *Let $D$ be an arbitrary domain, $\Gamma$ its boundary, $E$ a compact set of capacity zero on $\Gamma$ and $z_0$ a point of $E$ such that $U(z_0) \cap (\Gamma - E) \neq \theta$ for every neighborhood $U(z_0)$ of $z_0$. Let $w = f(z)$ be a single-valued meromorphic function in $D$ and bounded in the intersection of $D$ with some neighborhood of $z_0$. Then it holds*

$$\overline{\lim_{z \to z_0}}\, |f(z)| = \overline{\lim_{\substack{\zeta \to z_0 \\ \zeta \in \Gamma - E}}} \Big( \overline{\lim_{z \to \zeta}}\, |f(z)| \Big)\,. \tag{2}$$

*Furthermore, since the left-hand side and the right-hand side of* (2) *denote the radii $r(D)$ and $r(\Gamma - E)$ of the smallest closed circular discs with center at $w = 0$ which contain $C_D(f, z_0)$ and $C_{\Gamma - E}(f, z_0)$ respectively,* (2) *can be written in the form*

$$r(D) = r(\Gamma - E)\,. \tag{3}$$

Remark. In the special case where $z_0$ is a regular point for the Dirichlet problem, this theorem is an immediate corollary of Theorem 1, since $|f(z)|$ is subharmonic in $D \cap U(z_0)$ for some neighborhood $U(z_0)$.

The following two theorems are equivalent to Theorem 2.

**Theorem 3.** *If $\alpha$ does not belong to $C_D(f, z_0)$ (in place of the assumption in Theorem 2 that $w = f(z)$ is bounded in the intersection of $D$ with some neighborhood of $z_0$), then (2) can be replaced by*

$$\varrho(C_D(f, z_0), \alpha) = \varrho(C_{\Gamma-E}(f, z_0), \alpha), \tag{4}$$

*where $\varrho(S, \alpha)$ denotes the spherical distance of $\alpha$ from a set $S$.*

**Theorem 4.** *Let $D$ be an arbitrary domain, $\Gamma$ its boundary, $E$ a compact set of capacity zero on $\Gamma$ and $z_0$ a point of $E$ such that $U(z_0) \cap (\Gamma - E) \neq \theta$ for every neighborhood $U(z_0)$ of $z_0$. Let $w = f(z)$ be single-valued and meromorphic in $D$. Then*

$$\mathscr{F} C_D(f, z_0) \subset \mathscr{F} C_{\Gamma-E}(f, z_0), \tag{5}$$

*where $\mathscr{F} S$ denotes the frontier of a set $S$.*

Remark. Theorem 2 implies Theorem 3: Suppose $\alpha \notin C_D(f, z_0)$ and consider the function $W = F(z)$ obtained by composing a linear transformation $W = (1 + \bar{\alpha} w)/(w - \alpha)$ with $w = f(z)$. Then (3) holds for $W = F(z)$. In other words, the spherical distances of $W = \infty$ from $C_D(F, z_0)$ and $C_{\Gamma-E}(F, z_0)$ are identical. But, since the linear transformation is a rotation of the Riemann sphere, we have (4). Theorem 3 implies Theorem 4: Let $M$ and $N (N \subset M)$ be two closed sets in the $w$-plane. If $\varrho(M, P) = \varrho(N, P)$ for any point $P$ exterior to $M$, then $\mathscr{F} M \subset \mathscr{F} N$. A proof follows from this fact. That Theorem 4 implies Theorem 2 is obvious. Thus we see that

Theorem 2 → Theorem 3 → Theorem 4 → Theorem 2.

2. *Proof of Theorem 4.* Let $w_0$ be an arbitrary point belonging to $C_D(f, z_0) - C_{\Gamma-E}(f, z_0)$. By hypothesis, there exists a circle $K: |z - z_0| = r$, arbitrarily small, such that $K \cap E = \theta$ and $f(z) \neq w_0$ on $K \cap D$. We may assume that $w_0$ does not belong to the closure $M_r$ of the union $\bigcup_{\zeta} C_D(f, \zeta)$ for all $\zeta$ belonging to the intersection of $\Gamma - E$ with $|z - z_0| \leqq r$. We denote by $\varrho'$ the distance of $M_r$ from $w_0$. Let $\varrho''$ be a positive number such that $|f(z) - w_0| \geqq \varrho'' > 0$ on $K \cap D$. We denote by $\varrho$ a positive number less than $\min(\varrho', \varrho'')$. Since $w_0$ is a cluster value of $w = f(z)$ at $z_0$, there exists a sequence of points $z_\mu$ $(\mu = 1, 2, \ldots)$ inside $(K) \cap D$, $(K)$ denoting the interior of $K$, tending to $z_0$ such that $w_\mu = f(z_\mu)$ tends to $w_0$. We keep this sequence $z_\mu$ $(\mu = 1, 2, \ldots)$ throughout the proof. The inverse image $D_0$ of $(c): |w - w_0| < \varrho$ in $(K) \cap D$ consists of at most a countable number of connected components. The component containing $z_\mu$ is denoted by $\Delta_\mu$ (which may coincide with other $\Delta_\nu$).

We treat first the case where there is an infinite number of distinct components $\Delta_\mu$. In this case, we assume for the sake of simplicity that $\Delta_\mu \neq \Delta_\nu$ if $\mu \neq \nu$. Then $\Delta_\mu$ $(\mu = 1, 2, \ldots)$ converges to $z_0$. Otherwise, there will be a circle $K'$: $|z - z_0| = r' < r$ such that $K' \cap E = \theta$ and $K' \cap \Delta_{\mu_n} \neq \theta$ $(n = 1, 2, \ldots)$. Let $\zeta_n$ be a boundary point of $\Delta_{\mu_n}$ on the circle $K'$ and $\zeta_0$ an accumulation point of the sequence $\{\zeta_n\}$. Clearly $f(\zeta_n)$ lies on the circle $c$: $|w - w_0| = \varrho$. It is also clear that either $\zeta_0$ belongs to $\Gamma - E$ or $D$. We shall have a contradiction in either case, because either $M_r$ meets the circle $c$: $|w - w_0| = \varrho$ or else infinitely many of the level curves: $|f(z) - w_0| = \varrho$ meet a small neighborhood of $\zeta_0$ inside $D$. If $\Delta_\mu$ is compact in $D$[1], then $w = f(z)$ takes every value of $(c)$: $|w - w_0| < \varrho$. If $\Delta_\mu$ is not compact in $D$, then, by Lemma, § 3, the value-set $\mathfrak{D}_\mu = f(\Delta_\mu)$ is everywhere dense in $(c)$; i. e. the closure $\overline{\mathfrak{D}}_\mu$ coincides with the closed disc $\overline{(c)}$: $|w - w_0| \leqq \varrho$. Since $\Delta_\mu$ $(\mu = 1, 2, \ldots)$ converges to $z_0$, $C_D(f, z_0)$ includes the closed disc $\overline{(c)}$.

We consider two monotone decreasing sequences of positive numbers, $\{r_n\}$ and $\{\varrho_n\}$, converging to zero, such that, for each $n$, $r_n$ and $\varrho_n$ are selected as stated above, and two sequences of circles $K_n$: $|z - z_0| = r_n$ and $c_n$: $|w - w_0| = \varrho_n$ $(n = 1, 2, \ldots)$. We denote by $\Delta_\mu^{(n)}$ the connected component, containing $z_\mu$, of the inverse image of $(c_n)$: $|w - w_0| < \varrho_n$. Suppose that there exists at least one $n$ for which the sequence $\Delta_\mu^{(n)}$ $(\mu \geqq N(n))$ consists of infinitely many domains. We can conclude from the discussion above that $C_D(f, z_0)$ contains $|w - w_0| \leqq \varrho_n$. Thus it is sufficient to consider the case in which, for every $n$, $\Delta_\mu^{(n)}$ consists of a finite number of different domains. Denote by $\Delta^{(1)}$ any domain $\Delta_\mu^{(1)}$ containing a subsequence $\{z_\mu^{(1)}\}$ of $\{z_\mu\}$, and by $\Delta^{(2)}$ any domain $\Delta_\mu^{(2)}$ containing a subsequence $\{z_\mu^{(2)}\}$ of $\{z_\mu^{(1)}\}$ and so on. Thus we obtain a new sequence $\{\Delta^{(n)}\}$ such that $\Delta^{(1)} \supset \Delta^{(2)} \supset \cdots \supset \Delta^{(n)} \supset \cdots$; obviously all $\Delta^{(n)}$ have the boundary point $z_0$ in common. Since the value-set of $f(z)$ in $\Delta^{(n)}$ is included in $(c_n)$ and the diameter of $\Delta_n$ tends to zero as $n \to \infty$, it is shown that there exists a path $\Lambda$ in $D$ terminating at $z_0$ along which $w_0$ is an asymptotic value of $w = f(z)$ at $z_0$.

Now, we denote by $\Delta$ the component of $D_0$ (defined above) containing the last part of $\Lambda$. The boundary $\gamma$ of $\Delta$ consists of a closed subset $e$ of $E$ and at most a countable number of analytic curves (boundary of $\Delta$ relative to $D$). Suppose that $z_0$ is not a regular boundary point of $\Delta$ for the Dirichlet problem. Then, if we denote by $\Delta'$ the new domain obtained by cutting $\Delta$ along the last part of $\Lambda$, the boundary $\gamma'$ of $\Delta'$ consists of $\gamma$ and the last part of $\Lambda$. Since $z_0$ is now a regular boundary point of the domain $\Delta'$ for the Dirichlet problem, we can apply Theorem 4, as a corollary of Theorem 1 (see Remark of Theorem 2), to the domain $\Delta'$.

[1] This means that the closure of $\Delta_\mu$ lies completely inside $D$.

Thus we have

$$\mathscr{F} C_{\Delta'}(f, z_0) \subset \mathscr{F} C_{\gamma'-e}(f, z_0) . \tag{6}$$

It is clear that

$$C_\Delta(f, z_0) = C_{\Delta'}(f, z_0) ; \quad C_{\gamma'-e}(f, z_0) = C_{\gamma-e}(f, z_0) \cup (w_0)$$

and $C_{\gamma-e}(f, z_0)$ is a closed set on the circle $c$: $|w - w_0| = \varrho$. On the other hand, the cluster set $C_\Delta(f, z_0; \Lambda)$ (cf. § 1, I) is a continuum which is contained in $C_\Delta(f, z_0)$ and connects $w_0$ and $c$: $|w - w_0| = \varrho$. Consequently, every point $\alpha$ of $(c)$: $|w - w_0| < \varrho$ must belong to $C_\Delta(f, z_0)$. Contrary to this assertion, if $\alpha\,(\neq w_0)$ in $(c)$ does not belong to $C_\Delta(f, z_0)$, then there is a boundary point $\beta$ of $C_\Delta(f, z_0) = C_{\Delta'}(f, z_0)$ on the circle $|w - w_0| = |\alpha - w_0|$ so that $\beta$ cannot belong to $C_{\gamma'-e}(f, z_0)$. This contradicts the inclusion (6).

Thus we have proved that any point $w_0$ belonging to $C_D(f, z_0) - C_{\Gamma-E}(f, z_0)$ is an interior point of this set. Consequently $C_D(f, z_0) - C_{\Gamma-E}(f, z_0)$ is an open set; in other words, $\mathscr{F} C_D(f, z_0) \subset \mathscr{F} C_{\Gamma-E}(f, z_0)$.

*Back ground.* We consider the case where $E$ consists of a single point $z_0$. SEIDEL [1] proved Theorem 2, with an aid of Lindelöf's principle of maximum modulus, in the case of a Jordan domain $D$ and DOOB [1] proved the equivalence between Theorem 2 and Theorem 4 in the same case and obtained a simple proof for Iversen-Gross' theorem (cf. Paragraph 3). BEURLING [1] proved Theorem 4, by making use of his theory of harmonic majoration in the case where $z_0$ is a regular boundary point for the Dirichlet problem. KUNUGUI [1] found that the restriction of regularity is not necessary and gave a direct proof for it by using results of IRIE [1] and NOSHIRO [2]. BRELOT [1] and NOSHIRO [4] simplified Kunugui's proof. However, it was found later that this theorem had been proved essentially by IVERSEN [2]. The extension to the case where $E$ consists of a closed set of capacity zero was made by TSUJI [4].

**3.** We state an extension of Theorem 2, § 2.

**Theorem 5.** *If, under the same assumptions as in Theorem 4, the open set $\Omega = C_D(f, z_0) - C_{\Gamma-E}(f, z_0)$ is not empty, then every value of $\Omega$ is assumed by $w = f(z)$ infinitely often in any neighborhood of $z_0$ except for a possible set of values of capacity zero; i. e., $\Omega - R_D(f, z_0)$ is at most of capacity zero* (TSUJI [4]).

*Proof.* Let $e_n$ be the set of values, belonging to $\Omega$, taken by $w = f(z)$ exactly $n$ times $(n = 0, 1, \ldots)$ in the intersection $U(z_0) \cap D$ of $D$ with a neighborhood $U(z_0)$ of $z_0$. Clearly the set $e$ of values $w \in \Omega$ taken by $w = f(z)$ finitely often is the union $\bigcup_{n=0}^{\infty} e_n$ of $e_n$. If cap. $e > 0$, then there exists at least one $n$ such that cap. $e_n > 0$. We can find a point $w_0 \in e_n$ such that for any positive number $\varrho$, the intersection of $V(w_0, \varrho)$: $|w - w_0| < \varrho$ with $e_n$ is of positive capacity. Since there are just $n$ $w_0$-points of $w = f(z)$ in $U(z_0) \cap D$, for a sufficiently small $\varrho$, the inverse image of

$|w - w_0| < \varrho$ inside $U(z_0) \cap D$ contains $n$ islands (compact connected components) $\Delta_1, \Delta_2, \ldots, \Delta_n$ (clearly, if there are some multiple $w_0$-points, the number of the islands is less than $n$). Next we select a positive number $r$ such that these islands lie in the exterior of $K$: $|z - z_0| = r$ and such that

$$K \cap E = \emptyset, \quad f(z) \neq w_0 \quad \text{on} \quad K \cap D, \quad w_0 \notin M_r\text{.}^1$$

We denote by $\mathfrak{R}$ the set of values taken by $w = f(z)$ on $K \cap D$. Let $\varrho_0 (< \varrho)$ be a positive number less than the distance of $w_0$ from $\mathfrak{R} \cup M_r$. Since $w_0$ is a cluster value of $w = f(z)$ at $z_0$, $w = f(z)$ takes a value belonging to $(c)$: $|w - w_0| < \varrho_0$ at a point $z_1$ in $(K) \cap D$, where $(K)$ denotes the interior of $K$. We denote by $\Delta$ the component, containing the point $z_1$, of the inverse image of $(c)$. Clearly $\Delta$ is not compact in $D \cap (K)$. The boundary of $\Delta$ consists of some analytic curves and a closed subset $E_0$ of $E$ of capacity zero. We map by $z = z(\zeta)$ the universal covering surface $\tilde{\Delta}$ of $\Delta$ conformally onto the unit disc $|\zeta| < 1$ in a one to one manner. Denote by $E_\zeta$ the set of points $\zeta = e^{i\varphi}$ such that the radial limit $z(e^{i\varphi})$ of $z = z(\zeta)$ at $\zeta = e^{i\varphi}$ exists and $z(e^{i\varphi}) \in E_0$. Here we make use of Evans-Selberg's function $\chi(z)$ associated with $E_0$ [see (1), § 2]. The function $1/\chi(z(\zeta))$ is a non-constant bounded regular function in $|\zeta| < 1$ which has the radial limit zero at every point of $E_\zeta$. By Riesz' theorem, $E_\zeta$ must be of linear measure zero. On the other hand, since $(c) \cap e_n$ is of positive capacity, we can find a closed subset $e^*$ of $(c) \cap e_n$, lying completely inside $(c)$, of positive capacity. Denote by $\Delta_w$ the external connected component of the open set $(c) - e^*$ whose boundary consists of $c$: $|w - w_0| = \varrho_0$ and $\gamma_w \subset e^*$. Let $\omega(w) = \omega(w, \gamma_w, \Delta_w)$ be the harmonic measure for $\Delta_w$ with boundary value 0 on $c$. Then obviously $\omega(w)$ is not identically zero. Consider now the composed function $v(\zeta) = \omega(f(z(\zeta)))$ in $|\zeta| < 1$. Then $v(\zeta)$ is a bounded harmonic function in $|\zeta| < 1$ whose radial limit $v(e^{i\varphi})$ is zero for almost every $\varphi (0 \leqq \varphi < 2\pi)$; hence $v(\zeta)$ must be identically zero. Contradiction[2].

It is important to remark that if we impose some restrictions on the boundary $\Gamma$ of $D$, then the exceptional set in Theorem 5 becomes remarkably small.

As an extension of Beurling-Kunugui's theorem[3], we prove

**Theorem 6.** *Let $D$ be a domain, $\Gamma$ its boundary, $E$ a closed set of capacity zero contained in a single boundary component $\Gamma_0$ of $\Gamma$ and $z_0$*

[1] See the proof of Theorem 4.

[2] If we use the theory of open Riemann surfaces, the proof can be modified in the following way. Construct the double $\hat{\Delta}$, which is obtained from $\Delta$ by the process of symmetrization. Then, the abstract Riemann surface $\hat{\Delta}$ belongs to $O_G$, since the complement of $E_0$ with respect to the $z$-plane belongs to $O_G$ (see Kuroda [6], p. 234). It is also known that the covering surface $\hat{\Phi}$ conformally equivalent to $\hat{\Delta}$ covers every point in the $w$-plane except for a possible set of capacity zero. This contradicts the hypothesis that $(c) \cap e_n$ is of positive capacity.

[3] Beurling [1], Kunugui [2].

*a point of $E$ such that $U(z_0) \cap (\Gamma - E) \neq \theta$ for any neighborhood $U(z_0)$ of $z_0$. Suppose that $w = f(z)$ is single-valued and meromorphic in $D$ and that the open set $\Omega = C_D(f, z_0) - C_{\Gamma - E}(f, z_0)$ is not empty. Then, $w = f(z)$ takes every value, with two possible exceptions, belonging to any component $\Omega_n$ of $\Omega$, infinitely often in any neighborhood of $z_0$* (NOSHIRO [8]).

*Proof.* Without loss of generality, we may suppose that $\Omega_n$ does not contain $w = \infty$. Contrary to our assertion, we suppose that there are three exceptional values $w_0$, $w_1$ and $w_2$ in $\Omega_n$. Then, there exists a positive number $\eta_1$ such that $f(z) \neq w_0, w_1, w_2$ in the common part of $D$ and $U(z_0, \eta_1): |z - z_0| < \eta_1$. Inside $\Omega_n$ we draw a simple closed regular analytic curve $\mathscr{L}$ which encloses $w_0$, $w_1$ and passes through $w_2$[1], and whose interior consists of only interior points of $\Omega_n$. We can select a positive number $\eta\,(<\eta_1)$, arbitrarily small, such that $K \cap E = \theta$ and $K \cap \Gamma \neq \theta$, $K$ denoting the circle $|z - z_0| = \eta$, and such that the closure $M_\eta$ of the union $\bigcup_\zeta C_D(f, \zeta)$ for all $\zeta$ belonging to the intersection of $\Gamma - E$ with $|z - z_0| \leqq \eta$ lies outside $\mathscr{L}$. Now, by Theorem 1, § 3, $w_0$ is either an asymptotic value of $w = f(z)$ at $z_0$ or there exists a sequence $z_n' \in E$ tending to $z_0$ such that $w_0$ is an asymptotic value at each $z_n'$. Consequently, it is possible to find a point $z_0'$ (distict from $z_0$ or not) belonging to $E \cap U(z_0, \eta)$ such that $w_0$ is an asymptotic value of $w = f(z)$ at $z_0'$. Let $\Lambda$ be the asymptotic path with the asymptotic value $w_0$ at $z_0'$. We may assume that the image of $\Lambda$ by $w = f(z)$ is a curve lying completely inside $\mathscr{L}$. Consider the set $D_\eta$ of points $z$ in the intersection of $D$ with $U(z_0, \eta)$ such that $w = f(z)$ lies inside $\mathscr{L}$. Denote by $\Delta$ the component of $D_\eta$ which contains the path $\Lambda$. We shall show that $\Delta$ is simply-connected provided that $\eta$ is sufficiently small. As is easily seen, the boundary of $\Delta$ consists of a finite number of arcs on $K$, at most a countable number of analytic curves (relative boundary) inside $D$, and a closed subset $E_0$ of $E$. Note that the boundary of $\Delta$ does not contain any closed analytic curve (in $D$), since any analytic curve contained in the boundary of $\Delta$ is transformed by $w = f(z)$ into a curve lying on the simple closed curve $\mathscr{L}$ passing through the exceptional value $w_2$. Further, the boundary of the bounded domain $\Delta$ consists of a single continuum, since $E$ is contained in a single component $\Gamma_0$ of $\Gamma$. Thus it is concluded that $\Delta$ is simply connected. Next, we make use of Evans-Selberg's function $\zeta = \chi(z) = e^{u(z) + iv(z)} = \varrho(z)\, e^{iv(z)}$ associated with the closed set $E_0$ of capacity zero. Let $\Gamma_\lambda$ be the level curve $\varrho(z) = \text{const.} = \lambda$ $(0 < \lambda < \infty)$. As stated before, Evans-Selberg's function has an important property (2), § 2. Let $\lambda_0$ be a fixed positive number such that for $\lambda_0 \leqq \lambda$ all the level curves $\Gamma_\lambda$ intersect the asymptotic path $\Lambda$. For $\lambda_0 \leqq \lambda$, let $\Theta_\lambda$ denote the common part of the level curve $\Gamma_\lambda$ and the domain $\Delta$;

[1] This technic is due to TÔKI [1].

$\Theta_\lambda$ consists of only a finite number of cross-cuts and does not contain any loop-cut, as $\Delta$ is simply connected. Denote by $\Delta(\lambda)$ the common part of $\Delta$ and the domain exterior to $\Gamma_\lambda$. It is clear that the open set $\Delta(\lambda)$ consists of simply connected components. Let $A(\lambda)$ denote the area of the Riemannian image of the open set $\Delta(\lambda)$ by the function $w = f(z)$ and let $L(\lambda)$ denote the total length of the image of $\Theta_\lambda$. Then,

$$A(\lambda) = \iint_{\Delta(\lambda)} |f'(z)|^2 \, d\tau \quad (d\tau\text{: the area element in the } z\text{-plane}),$$

$$L(\lambda) = \int_{\Theta_\lambda} |f'(z)| \, |dz| \, .$$

Next we shall prove that

$$\varliminf_{\lambda \to \infty} A(\lambda) = \infty \tag{7}$$

and

$$\varliminf_{\lambda \to \infty} L(\lambda)/S(\lambda) = 0 \quad \text{where}$$

$$S(\lambda) = \frac{A(\lambda)}{\text{area of the interior of } \mathscr{L}} \, . \tag{8}$$

To prove these, we use a branch of Evans-Selberg's function

$$\zeta = \chi(z) = e^{u(z) + iv(z)} \, , \quad (0 \leqq v(z) < 2\pi)^1 \, . \tag{9}$$

By setting $W(\zeta) = f[\chi^{-1}(\zeta)]$, we have

$$A(\lambda) - A(\lambda_0) = \int_{\lambda_0}^{\lambda} \int_{\tilde{\Theta}_\lambda} |W'(\zeta)|^2 \lambda \, d\lambda \, d\theta \, , \quad (\zeta = \lambda e^{i\theta}) \, , \tag{10}$$

where $\tilde{\Theta}_\lambda$ denotes the image of $\Theta_\lambda$ on the circle $|\zeta| = \lambda$ under the transformation $\zeta = \chi(z)$ $(0 \leqq v(z) < 2\pi)$, and

$$L(\lambda) = \int_{\tilde{\Theta}_\lambda} |W'(\zeta)| \, \lambda \, d\theta \, .$$

Denote by $\sigma > 0$ the distance of $\mathscr{L}$ from the image of $\Lambda$. Then, a geometric consideration gives $L(\lambda) \geqq 2\sigma$ for $\lambda_0 \leqq \lambda < \infty$. Applying Schwarz's inequality, we have

$$[L(\lambda)]^2 \leqq \int_{\tilde{\Theta}_\lambda} \lambda \, d\theta \int_{\tilde{\Theta}_\lambda} |W'(\zeta)|^2 \lambda \, d\theta = \lambda \, \theta(\lambda) \int_{\tilde{\Theta}_\lambda} |W'(\zeta)|^2 \lambda \, d\theta \, ,$$

i. e.

$$\frac{[L(\lambda)]^2}{\lambda \, \theta(\lambda)} \leqq \int_{\tilde{\Theta}_\lambda} |W'(\zeta)|^2 \lambda \, d\theta \, . \tag{11}$$

Consequently

$$\frac{2\sigma^2}{\pi} \int_{\lambda_0}^{\lambda} \frac{d\lambda}{\lambda} \leqq \int_{\lambda_0}^{\lambda} \int_{\tilde{\Theta}_\lambda} |W'(\zeta)|^2 \lambda \, d\lambda \, d\theta = A(\lambda) - A(\lambda_0) \, , \tag{12}$$

since

$$\theta(\lambda) = \int_{\Theta_\lambda} dv(z) \leqq \int_{\Gamma_\lambda} dv(z) = 2\pi \, . \tag{13}$$

[1] See the foot-note 1 of § 2, p. 13.

Inequality (12) yields (7) when $\lambda$ tends to infinity. Next we obtain from (11)

$$\frac{d\lambda}{\lambda\,\theta(\lambda)} \leqq \frac{dA(\lambda)}{[L(\lambda)]^2}.$$

Hence, denoting by $M_\lambda$ the set of all $\lambda$ such that

$$L(\lambda) \geqq A(\lambda)^{\frac{1}{2}+\varepsilon}, \quad (\varepsilon > 0),$$

we see, by (13), that

$$\frac{1}{2\pi}\int\limits_{M_\lambda} d\log\lambda \leqq \int\limits_{M_\lambda} \frac{d\lambda}{\lambda\,\theta(\lambda)} \leqq \int\limits_{M_\lambda} \frac{dA(\lambda)}{[A(\lambda)^{\frac{1}{2}+\varepsilon}]^2} \leqq \int^{\infty} \frac{dt}{t^{1+2\varepsilon}} < \infty,$$

whence $L(\lambda) < A(\lambda)^{\frac{1}{2}+\varepsilon}$ for all $\lambda$ not belonging to a set $M_\lambda$, where $\int\limits_{M_\lambda} d\log\lambda < \infty$. Thus (8) follows.

If $\lambda_0 \leqq \lambda$, the open set $\Delta(\lambda)$ consists of a finite number of simply connected components which we shall denote by

$$\Delta^{(1)}(\lambda), \quad \Delta^{(2)}(\lambda), \ldots, \quad \Delta^{(m)}(\lambda),$$

where $m = m(\lambda)$, $m \geqq 1$, depends on $\lambda$. Denote by $\Phi^{(i)}(\lambda)$ the Riemannian image of $\Delta^{(i)}(\lambda)$ under $w = f(z)$ $(i = 1, 2, \ldots, m)$. If we denote by $\Phi_0$ the domain obtained by excluding the two points $w_0$ and $w_1$ from the interior of $\mathscr{L}$, then, by hypothesis, $\Phi^{(i)}(\lambda)$ $(i = 1, 2, \ldots, m)$ is a finite covering surface of the basic surface $\Phi_0$. By Ahlfors' principal theorem on covering surfaces[1], we have

$$S^{(i)} \leqq h\,L^{(i)} \qquad (i = 1, 2, \ldots, m), \tag{14}$$

where $S^{(i)}$ denotes the average number of sheets $\Phi^{(i)}(\lambda)$, i. e., $S^{(i)}$ denotes the ratio between the area of $\Phi^{(i)}(\lambda)$ and the area of $\Phi_0$, and $L^{(i)}$ the length of the boundary of $\Phi^{(i)}(\lambda)$ relative to $\Phi_0$, $h$ being a constant dependent only on $\Phi_0$. From (14)

$$\sum_{i=1}^{m} S^{(i)} \leqq h \sum_{i=1}^{m} L^{(i)},$$

that is,

$$S(\lambda) \leqq h(L(\lambda) + L_0), \tag{15}$$

where $L_0$ denotes the total length of the image of the arcs of $K$ included in the boundary of $\Delta$. Accordingly

$$\lim_{\lambda\to\infty} \frac{L(\lambda)}{S(\lambda)} \geqq \frac{1}{h} > 0. \tag{16}$$

It is clear that (16) contradicts (8), so that our theorem is proved.

In the special case where $E$ consists of a single point $z_0$, Theorem 6 becomes

**Theorem 7.** (Beurling-Kunugui's theorem). *Let $D$ be an arbitrary domain, $\Gamma$ its boundary and $z_0$ a non-isolated boundary point. Let $w = f(z)$*

[1] Ahlfors [2].

*be single-valued and meromorphic in D. Suppose that* $\Omega = C_D(f, z_0) - C_\Gamma(f, z_0)$ *is not empty, and let* $\Omega_n$ *be any connected component of* $\Omega$. *Then, every value of* $\Omega_n$ *is assumed by* $w = f(z)$, *except for at most two values, infinitely often in any neighborhood of* $z_0$.

This theorem is a generalization of the following classical

**Theorem of Iversen-Gross**[1]. *Let D be a domain bounded by a closed Jordan curve* $\Gamma$ *and* $z_0$ *be a boundary point. Let* $w = f(z)$ *be meromorphic in D. Suppose that* $\Omega = C_D(f, z_0) - C_\Gamma(f, z_0)$ *is not empty. Then, every value of* $\Omega$ *belongs to* $R_D(f, z_0)$ *with two possible exceptions.*

*Proof.* Let $\alpha_1$ and $\alpha_2$ be two exceptional values, in a neighborhood of $z_0$, belonging to $\Omega$. Then, by Theorem 1, § 3, $\alpha_1$ and $\alpha_2$ are asymptotic values of $f(z)$ along two paths $\Lambda_1$ and $\Lambda_2$ in $D$ terminating at $z_0$. By a well-known theorem of Iversen-Lindelöf[2] on asymptotic values, $w = f(z)$ takes every value infinitely often except perhaps two values between $\Lambda_1$ and $\Lambda_2$. Accordingly, $R_D(f, z_0)$ coincides with the domain obtained by excluding two points $\alpha_1$ and $\alpha_2$ from the $w$-plane; hence $C_D(f, z_0)$ is the whole $w$-plane. As an alternative, if $C_D(f, z_0)$ is not the whole $w$-plane, every value of $\Omega$ belongs to $R_D(f, z_0)$ with one possible exception.

*Back ground.* Doob [1] gave a simple proof for Theorem 7 in the case of a Jordan domain, applying Iversen-Lindelöf's theorem on asymptotic values. Beurling [1] conjectured that every value of $\Omega$ in Theorem 7 will be assumed infinitely often by $f(z)$ in any neighborhood of $z_0$ except for at most two values of $\Omega$. Kunugui [2] constructed a counter-example to this conjecture. Kunugui's example with a slight modification is as follows. Starting from a functional element of $g(z) = \sqrt{e^{z^2} - 1} = z + \frac{1}{4} z^3 + \cdots$, we construct a star domain $D$ in the sense of Gross and denote by $\Gamma$ its boundary. Then $\Gamma$ consists of four rays: $|z| \geqq \sqrt{2\pi}$, $\arg z = \frac{1}{4}(2k+1)\pi$ $(k = 0, 1, 2, 3)$. Let $z_0 = \infty$. Then it is obvious that $C_D(g, z_0)$ is the whole $w$-plane, $C_\Gamma(g, z_0)$ is the lemniscate: $|w^2 + 1| = 1$, and $C_D(g, z_0) - C_\Gamma(g, z_0)$ consists of three connected components. The function $g(z)$ has three exceptional values $-i$, $i$, $\infty$. Note that $z_0 = \infty$ is a regular boundary point for the Dirichlet problem. Kunugui proved that if one takes $\Omega_n$ instead of $\Omega$, then Beurling's conjecture holds, and investigated the conditions to be placed on $\Gamma$ in order that the original conjecture of Beurling holds.

As an immediate corollary of Theorem 6, we have

**Theorem 8.** *Let D be a simply connected domain of hyperbolic type, E a closed set of capacity zero contained in the boundary* $\Gamma$, *and* $z_0$ *a point of E. Let* $w = f(z)$ *be single-valued and meromorphic in D. Let* $\Omega = C_D(f, z_0) -$

[1] Iversen [2], Gross [2], Doob [1].

[2] Iversen [1].

*— $C_{\Gamma - E}(f, z_0)$ be non-empty and $\Omega_n$ be any connected component of $\Omega$. Then, $w = f(z)$ takes every value, with two possible exceptions, belonging to $\Omega_n$ infinitely often in any neighborhood of $z_0$.*

Furthermore we have

**Theorem 9.** *Let D be a simply connected domain of hyperbolic type, E a closed set of capacity zero contained in the boundary $\Gamma$, and $z_0$ a point of E. Let $w = f(z)$ be single-valued and meromorphic in D. Suppose that $\Omega = C_D(f, z_0) - C_{\Gamma - E}(f, z_0) \neq \emptyset$ and that $f(z)$ is regular in the common part of D and a certain neighborhood of $z_0$. Let $\Omega_n$ be any connected component of $\Omega$. Then $w = f(z)$ takes every finite value, with one possible exception, belonging to $\Omega_n$ infinitely often in any neighborhood of $z_0$*[1].

*Proof.* Suppose that there are two finite exceptional values $w_0$ and $w_1$ within $\Omega_n$, and let $\mathscr{L}$ be any simple closed regular analytic curve, in $\Omega_n$, which surrounds $w_0$ and $w_1$ and whose interior consists of only interior points of $\Omega_n$. Let $\Delta$ be the domain defined in the same way as in the proof of Theorem 6. Thus, we can easily see that $\Delta$ is also simply connected, for if $\Delta$ were not simply connected, the boundary of $\Delta$ would contain at least one closed analytic contour $q$ such that $q$ is a loop-cut of $D$. Accordingly $w = f(z)$ would take inside $q$ a value lying outside the simple closed curve $\mathscr{L}$, while $w = f(z)$ is regular both inside and on $q$, and the image of $q$ by $w = f(z)$ would lie on $\mathscr{L}$. This is a contradiction. Repeating the same argument as in the proof of Theorem 6, we arrive at a contradiction.

As an immediate consequence of Theorem 9, we see that under the same conditions as in Theorem 9, for any connected component $\Omega_n$ which does not contain $w = \infty$, $w = f(z)$ takes every value, with one possible exception, belonging to $\Omega_n$ infinitely often near $z_0$.

Thus we have

**Theorem 10.** *Let D be a simply connected domain of hyperbolic type, let $\Omega = C_D(f, z_0) - C_{\Gamma - E}(f, z_0)$ not be empty, and let $f(z)$ be regular and bounded in the common part of D and a certain neighborhood $U(z_0)$ of $z_0$ (or else $C_D(f, z_0)$ not coincide with the whole w-plane). Let $\Omega_n$ be any connected component of $\Omega$. Then $w = f(z)$ takes every value, with one possible exception, belonging to $\Omega_n$ infinitely often in any neighborhood of $z_0$.*

As another immediate consequence of Theorem 9, we obtain, by using a linear transformation,

**Theorem 11.** *Under the same conditions as in Theorem 8, if there are two exceptional values $w_0$, $w_1$ ($w_0 \neq w_1$) belonging to the same component $\Omega_n$, $w = f(z)$ takes every value other than $w_0$ and $w_1$ infinitely often in any neighborhood of $z_0$, so that $C_D(f, z_0)$ coincides with the whole w-plane.*

Here we state an application of Theorem 10 to the theory of conformal mappings.

[1] Noshiro [7].

**Theorem 12.** *Let e be a closed set of capacity zero, lying completely inside the unit circle (c): $|w| < 1$ and let $\Phi$ denote the domain obtained by excluding the set e from the disc (c). Let $w = f(z)$ be a function which maps the unit disc $|z| < 1$ conformally onto the universal covering surface $\tilde{\Phi}$ of $\Phi$ in a one-to-one manner. Suppose further that e contains at least two points. Then the perfect set E, on $|z| = 1$, of essential singularities of $w = f(z)$ must be of linear measure zero but the capacity of E must be positive* (P. J. MYRBERG [3])[1].

*Proof*[2]. Since $w = f(z)$ is regular and bounded: $|f(z)| < 1$ in $D$: $|z| < 1$, by Fatou's theorem, $w = f(z)$ has a radial limit $f(e^{i\theta})$ at almost every $z = e^{i\theta}$. Considering the function $1/\chi(f(z))$, where $\chi(w)$ denotes an Evans-Selberg's function associated with the set $e$, we can easily show that the set of values $\theta$ $(0 \leqq \theta \leqq 2\pi)$ such that $|f(e^{i\theta})| = 1$ is of linear measure $2\pi$.[3] We shall now investigate the set of singularities of $w = f(z)$ on the circumference $\Gamma$: $|z| = 1$. Since $w = f(z)$ is an automorphic function with a Fuchsian group $G$, it follows that the set $E_1$ corresponding to the circle $|w| = 1$ consists of an infinite number of open arcs $(a_n, b_n)$ $(n = 1, 2, \ldots)$, the two end-points $a_n$ and $b_n$ being fixed points of a hyperbolic linear transformation belonging to the group $G$, so that the complement $E$ of $E_1$ with respect to $|z| = 1$ is a closed set which consists of only singular points of $w = f(z)$. Remark that $E$ is a perfect set[4]. Suppose now that $E$ is of capacity zero. Let $z_0$ be any point of $E$. Then, it is clear that $C_D(f, z_0)$ is the closed disc $|w| \leqq 1$, $C_{\Gamma - E}(f, z_0)$ is the circumference $|w| = 1$ and $\Omega = C_D(f, z_0) - C_{\Gamma - E}(f, z_0)$ is the unit disc $|w| < 1$. Theorem 10 shows that $w = f(z)$ takes every value, with one possible exception, belonging to (c): $|w| < 1$ infinitely often in any neighborhood of $z_0$. Thus we arrive at a contradiction.

## § 5. Hervé's theorems

**1.** Let $D$ be an arbitrary domain with boundary $\Gamma$. Let $E$ be a compact set of capacity zero lying on $\Gamma$. Let $w = f(z)$ be single-valued and meromorphic in $D$. First, consider the case in which, for a point $\zeta \in E$, $C_{\Gamma - E}(f, \zeta) = \theta$, i. e. $\zeta \notin \overline{(\Gamma - E)}$ and $C_D(f, \zeta)$ consists of a single point; clearly $C_D(f, \zeta) - C_{\Gamma - E}(f, \zeta)$ is not open. Take a neighborhood $V(\zeta)$ of $\zeta$ sufficiently small so that $V(\zeta) \cap (\Gamma - E) = \theta$ and describe a simple closed curve $L$ around $\zeta$ inside $V(\zeta)$ which does not meet $E$. Denote by $e$ the subset of $E$ inside $L$ and by $D'$ the domain obtained by deleting $e$

[1] Compare this theorem with a result of L. MYRBERG [4].

[2] NOSHIRO [6].

[3] I. e., $w = f(z)$ is a function of class $(U)$ in the sense of SEIDEL; cf. § 1, III.

[4] For if there were an isolated point $z_0$ of $E$, then, by Schwarz's principle of reflection, $w = f(z)$ would be continued to a single-valued regular function in a domain $0 < |z - z_0| < r$, for a sufficiently small $r$, which omits at least four values; this contradicts a well-known Picard's theorem (cf. § 2, I).

from the interior of $L$. Then, the value-set of $w = f(z)$ in $D'$ is not dense in the $w$-plane. Hence, the set $e$ contains no transcendental singularity by Nevanlinna's theorem (cf. Theorem 1, § 2); i. e., $w = f(z)$ can be continued to a single-valued meromorphic function in the interior of $L$. If we denote by $E_1$ the set of $\zeta \in E$ for which $C_{\Gamma - E}(f, \zeta) = \theta$ and $C_D(f, \zeta)$ consists of a single point, then $E_1$ is relatively open in $E$ and then $w = f(z)$ is single-valued and meromorphic in $D \cup E_1$.

Next, suppose that $C_{\Gamma - E}(f, z_0) \neq \theta$ for a point $z_0$ of $E$ and that the open set $\Omega = C_D(f, z_0) - C_{\Gamma - E}(f, z_0)$ is not empty. Denote by $\Omega_n$ be a connected component of $\Omega$. Let $G$ be a domain whose closure $\overline{G}$ lies in $\Omega_n$. We can find a positive number $r$, sufficiently small, such that, if $|\zeta - z_0| \leqq r$ and if $\zeta \in \Gamma - E$, $C_D(f, \zeta) \cap \overline{G} = \theta$. We prove that all the points $\zeta$ of $E$ in the circle $|z - z_0| < r$ are classified into the following three kinds:

(i) $\zeta \in E_1$, (ii) $\overline{G} \subset C_D(f, \zeta)$; the set of all such points $\zeta$ will be denoted by $E(G)$, (iii) $G \cap C_D(f, \zeta) = \theta$. For this purpose, it is sufficient to show that if $\zeta \notin E_1$, then either (ii) or (iii) must hold. Assume, contrary to the assertion, that $\zeta \notin E_1$, $G \cap C_D(f, \zeta) \neq \theta$ and $G \not\subset C_D(f, \zeta)$. Then, obviously, $G \cap \mathscr{F} C_D(f, \zeta) \neq \theta$. On the other hand, by Theorem 4, § 4, we have $\mathscr{F} C_D(f, \zeta) \subset \mathscr{F} C_{\Gamma - E}(f, \zeta)$. Consequently, we would have $G \cap \mathscr{F} C_{\Gamma - E}(f, \zeta) \neq \theta$ which contradicts our choice of the positive number $r$.

Now, assume that $z_0 \in E$ is not an accumulation point of $E_1$. Let $G_1$ and $G_2$ be two domains such that $\overline{G}_1$ and $\overline{G}_2$ are contained in a component $\Omega_n$ of $\Omega$. Let $G$ be a domain such that $G$ contains both $\overline{G}_1$ and $\overline{G}_2$ in its interior and such that $\overline{G} \subset \Omega_n$. We choose a positive number $r$ such that $C_D(f, \zeta) \cap \overline{G} = \theta$ if $|\zeta - z_0| \leqq r$, $\zeta \in \Gamma - E$. We assume moreover that there exists no point of $E_1$ in $|z - z_0| \leqq r$. Clearly $E(G) \subset E(G_1)$. If $\zeta \in E(G_1)$, then $\overline{G}_1 \subset C_D(f, \zeta)$. Hence $C_D(f, \zeta) \cap G \neq \theta$. Accordingly, it follows that $\overline{G} \subset C_D(f, \zeta)$ and $\zeta \in E(G)$. Thus $E(G) = E(G_1)$. Similarly $E(G) = E(G_2)$. Thus $E(G_1) = E(G_2)$.

As an immediate consequence of this fact, if $z_0$ is not an accumulation point of $E_1$ and if $C_\Gamma(f, z_0) - C_{\Gamma - E}(f, z_0) \neq \theta$, then this set is the union of some components $\Omega_n$ of $\Omega = C_D(f, z_0) - C_{\Gamma - E}(f, z_0)$ and hence an open set. For if $w \in C_\Gamma(f, z_0) - C_{\Gamma - E}(f, z_0)$, then any domain $G$ which contains $w$ in its interior and whose closure is contained in $\Omega_n$, must be contained in $C_\Gamma(f, z_0) - C_{\Gamma - E}(f, z_0)$.

**Theorem 1.** *Let $D$ be a domain, $\Gamma$ its boundary, $E$ a countable compact set contained in $\Gamma$ and $z_0$ a point of $E$. Suppose that $f(z)$ is single-valued and meromorphic in $D$ and that $C_{\Gamma - E}(f, z_0) \neq \theta$ and $\Omega = C_D(f, z_0) - C_{\Gamma - E}(f, z_0) \neq \theta$. If $z_0$ is not an accumulation point of $E_1$, then every value of each component $\Omega_n$ of $\Omega$ is assumed by $w = f(z)$ infinitely often except for at most two values in any neighborhood of $z_0$. If $z_0$ is an accumulation point of $E_1$, then the set of exceptional values in the same sense is at most countable* (Hervé [1]).

*Proof*. Consider the case in which $z_0 \notin E_1'$, where $E_1'$ denotes the derived set of $E_1$. Suppose that there are three exceptional values $w_0, w_1, w_2$ in a component $\Omega_n$ of $\Omega$. As usual, in $\Omega_n$ we draw a simple closed regular analytic curve $\mathscr{L}$ enclosing $w_0$ and $w_1$ and passing through $w_2$ such that the interior $G$ of $\mathscr{L}$ consists of only points of $\Omega_n$. Select a positive number $r$ sufficiently small such that $C_D(f, \zeta) \cap \overline{G} = \theta$ if $|\zeta - z_0| \leqq r$ and if $\zeta \in \Gamma - E$ and such that the intersection of $|z - z_0| \leqq r$ with $D$ contains no point of $E_1$, and such that $f(z) \neq w_0, w_1, w_2$ in this intersection. If $E$ consists of a single point $z_0$, the theorem is already proved (cf. Theorem 7, § 4). The points $\zeta$ of $E$ inside $|z - z_0| < r$ satisfy (ii) or (iii). The set $E(G)$ of all the points satisfying $\overline{G} \subset C_D(f, \zeta)$ is clearly a closed set. Since $E$ is countable, $E(G)$ can not be a perfect set and, hence, $E(G)$ contains an isolated point $\zeta_0$. Then, there exists a path $\Lambda$ terminating at $\zeta_0$ along which $w_0$ is an asymptotic value of $w = f(z)$ at $\zeta_0$ by Theorem 1, § 3. Consider the inverse image $f^{-1}(G)$ of $G$ in $D$ by $w = f(z)$ and denote by $\Delta$ the component of the intersection of $f^{-1}(G)$ with a circular disc $|z - \zeta_0| < \varrho$ containing the last part of $\Lambda$. If $\varrho$ is chosen to be suitably small, $\Delta$ is a simply connected domain bounded by analytic curves in $D$, the point $\zeta_0$ and some circular arcs of $|z - \zeta_0| = \varrho$. The argument used before (cf. Proof of Theorem 6, § 4) shows that this leads to a contradiction. Next consider the case that $z_0 \in E_1'$.[1] Set $D_1 = D \cup E_1$ and $\Gamma_1 = \Gamma - E_1$. Clearly $C_D(f, z_0) = C_{D_1}(f, z_0)$ and $C_{\Gamma - E}(f, z_0) = C_{\Gamma_1 - (E - E_1)}(f, z_0)$. If we consider the function $f(z)$ in $D_1$, then the above discussion is available. Every value of each component $\Omega_n$ of $\Omega$ is assumed infinitely often by $w = f(z)$ in the intersection of a neighborhood of $z_0$ with $D_1$ except for at most two values. Since $D$ is obtained by excluding a countable set of points from $D_1$, every value of each component $\Omega_n$ is assumed infinitely often by $w = f(z)$ with a possible exceptional countable set of values.

2. HERVÉ has generalized Theorem 6, § 4 in the following form:

**Theorem 2.** *Let $D$ be a domain, $\Gamma$ its boundary, $E$ a compact set of capacity zero contained in $\Gamma$, such that each point of $E$ belongs to a boundary component of $D$ consisting of a non-degenerate continuum, and $z_0$ a point of $E$. Suppose that $f(z)$ is single-valued and meromorphic in $D$ and that the open set $\Omega = C_D(f, z_0) - C_{\Gamma - E}(f, z_0)$ is non-empty. Then, $w = f(z)$ takes every value, with two possible exceptions, belonging to each component $\Omega_n$ of $\Omega$, infinitely often in any neighborhood of $z_0$* (HERVÉ [2]).

For the proof, it seems necessary to have some preliminary discussions. Suppose that a connected component $\Omega_n$ of $\Omega$ contains three values $w, w', w^* \notin R_D(f, z_0)$. Let $U$ be a simply connected domain which contains the three points $w, w', w^*$ and whose closure $\overline{U}$ is situated in $\Omega_n$.

[1] Obviously, $z_0$ does not belong to $E_1$.

We can find a positive number $r_0$ such that $f(z) \neq w, w', w^*$ for $|z - z_0| \leqq r_0$, $z \in D$ and such that $\overline{U} \cap \overline{\bigcup_\zeta C_D(f, \zeta)} = \theta$ for $|\zeta - z_0| \leqq r_0$, $\zeta \in \Gamma - E$. Then, it is clear that

$$\overline{U} \subset C_D(f, \zeta) \quad \text{or} \quad U \cap C_D(f, \zeta) = \theta$$

for $|\zeta - z_0| < r_0$, $\zeta \in E$. We denote by $E(U)$ the set of all the points $\zeta$ such that $\overline{U} \subset C_D(f, \zeta)$. Inside $U$, we first describe a simple closed analytic curve $\mathcal{L}$ such that $w'$ and $w^*$ lie in its interior and $w$ lies in its exterior. Next, we describe another simple closed analytic curve $\mathcal{L}'$ in the interior of $\mathcal{L}$ such that $w'$ lies in the interior of $\mathcal{L}'$ and $w^*$ in the exterior of $\mathcal{L}'$.

**Lemma.** *If $L$ is a simple closed curve, described in the closed disc $|z - z_0| \leqq r_0$, which does not pass through any point of $E$ and contains at least one point of $E(U)$ in its interior, then the open set $G$ of all the points $z$ in the interior of $L$ such that $z \in D$, $f(z) \in (\mathcal{L}, \mathcal{L}')$, where $(\mathcal{L}, \mathcal{L}')$ denotes a ring-domain bounded by $\mathcal{L}$ and $\mathcal{L}'$, contains at least one component $\Delta$ with infinite connectivity* (Hervé [2]).

*Proof.* For every $\zeta \in E(U)$ in the interior of $L$,

$$w^* \in C_D(f, \zeta) - C_{\Gamma - E}(f, \zeta) - R_D(f, \zeta) .$$

Accordingly, there exists a path $\Lambda$ inside $L$, terminating at a point $\zeta'$ of $E$ arbitrarily near $\zeta$, along which $f(z)$ has an asymptotic value $w^*$ at $\zeta'$. We may assume that the image of $\Lambda$ lies in the interior of $(\mathcal{L}, \mathcal{L}')$. Suppose that the connected component $\Delta$ of $G$ containing the path $\Lambda$ has finite connectivity. Let $\chi_1, \chi_2, \ldots, \chi_q$ be the complementary continua of $\Delta$ other than the complementary continuum containing $z = \infty$. We easily see that any $\chi_i$ cannot degenerate to a single point of $E$. For, if $\chi_j$ be a single point of $E$, then we would have $C_D(f, \chi_j) \subset \overline{(\mathcal{L}, \mathcal{L}')}$; this contradicts the fact that

$$C_D(f, \chi_j) \supset \overline{U} \quad \text{or} \quad C_D(f, \chi_j) \cap U = \theta .$$

Accordingly, we can find a point $z_j \notin E$ in each $\chi_j$. Now, consider an Evans-Selberg's potential $u(z)$ of positive mass-distribution $\mu$ of total mass unity on $E$ (cf. Theorem 1, § 1) and set

$$\max_{1 \leqq j \leqq q} u(z_j) = m .$$

By the minimum principle of harmonic functions, we see that if $\beta$ is a simple closed curve in $\Delta$ such that $u(z) > m$ on $\beta$, then the interior of $\beta$ is contained in $\Delta$. Consequently, the open set of all the points $z$ of $\Delta$ satisfying $u(z) > m$ consists of only simply connected components. On the other hand, $(\mathcal{L}, \mathcal{L}') - w^*$ is triply connected and $f(z) \in (\mathcal{L}, \mathcal{L}') - w^*$ in these simply connected components. Considering the simply connected component containing the last part of $\Lambda$, we shall arrive at a contradiction[1].

3. *Proof of Theorem 2*[2]. Assume that there are three exceptional values $w$, $w'$, $w^*$ in $\Omega_n$. We take as $L$ in the above Lemma a circumference with center $z_0$. In the set $G$ in the Lemma, we can draw two simple closed curves $L_1^1$, $L_1^2$ exterior to each other whose interiors are not contained in $G$. If $a \in D$ is a boundary point of $G$ interior to $L_1^j$ ($j = 1$ or $2$), then $f(a) \in \mathscr{L}$ or $f(a) \in \mathscr{L}'$. If $f(a) \in \mathscr{L}$ (resp. $\mathscr{L}'$), we join the point $w$ (resp. $w'$) with $f(a)$ by an arc $\gamma$ described in $U$ and exterior to $\mathscr{L}$ (resp. interior to $\mathscr{L}'$) and consider the inverse image $f^{-1}(\gamma)$ of $\gamma$. Then, on the arc $f^{-1}(\gamma)$ there exists a sequence of points converging to a point of $E(U)$. Thus, we see that each $L_1^j$ contains at least one point of $E(U)$ in its interior. Accordingly, we can apply the Lemma by replacing $\mathscr{L}$ by another simple closed analytic curve $\mathscr{L}_1$ in the interior of $\mathscr{L}$, such that $\mathscr{L}_1$ contains $w^*$ and $\mathscr{L}'$ in its interior, and $\mathscr{L}'$ by another simple closed analytic curve $\mathscr{L}_1'$ in the interior of $\mathscr{L}_1$, such that $w^*$ lies in the exterior of $\mathscr{L}_1'$ and $\mathscr{L}'$ is interior to $\mathscr{L}_1'$. Then, in the open set $G_1^j$ of points $z$ interior to $L_1^j$ such that $z \in D$, $f(z) \in (\mathscr{L}_1, \mathscr{L}_1')$, we can describe two simple closed curves $L_2^{j1}$, $L_2^{j2}$, exterior to each other, whose interiors are not contained in $G_1^j$. Repeating this procedure indefinitely, we can choose two sequences of curves $\mathscr{L}_n$, $\mathscr{L}_n'$ ($n = 1, 2, \ldots$) in such a manner that they have a common limiting curve $\mathscr{L}^*$ passing through the point $w^*$. Denote by $K$ (resp. $K_1^j$, $K_2^{j1}$, ...) the continuum formed by the union of $L$ (resp. $L_1^j$, $L_2^{j1}$, ...) and its interior. For each sequence of integers $j_n$ ($= 1$ or $2$), the continua $K_n^{j_1, j_2, \ldots, j_n}$ are decreasing and have a common (non-empty) continuum $K^*$. Let $c$ be a boundary point of $K^*$. Then clearly $c \in E(U)$ or $c \in D$ and $f(c) \in \mathscr{L}^*$. But, neither the boundary of $K^*$ can degenerate to a single point of $E(U)$ (by the hypothesis of the theorem) nor to a point of $D$ because $K_n^{j_1, j_2, \ldots, j_n}$ contains at least one point of the closed set $E$. Thus, on the common part of the boundary of $K^*$ with $D$, $f(z)$ assumes all the values belonging to a certain non-degenerate arc of $\mathscr{L}^*$. It is clear that the set of sequences $\{K_n^{j_1, j_2, \ldots, j_n}\}$ has the power $2^{\aleph_0}$. Thus, it is concluded that $f(z)$ assumes a common value belonging to $\mathscr{L}^*$ on a non-enumerable set of different points in $D$. This is a contradiction.

4. HERVÉ [3] has sharpened Theorem 1 and Theorem 2.

Let $D$ be an arbitrary domain, $\Gamma$ the boundary of $D$, $E$ a compact set of capacity zero contained in $\Gamma$ and $w = f(z)$ a single-valued meromorphic function in $D$. The cluster set $C_\Delta(f, z_0)$ of $w = f(z)$ at $z_0 \in \Gamma$ is defined as follows: $\alpha \in C_\Delta(f, z_0)$ if there exists a sequence of points $z_n$ in $\Delta$ such that $z_0 = \lim_{n\to\infty} z_n$ and $\alpha = \lim_{n\to\infty} f(z_n)$, where $\Delta$ is an arbitrary set

[1] Cf. Proof of Theorem 6, § 4.

[2] This proof is due to HERVÉ [2].

such that $\Delta \subset D$ and $z_0 \in \bar{\Delta}$. If $f(z) \neq \alpha$ in the intersection of $D$ with some neighborhood of $z_0$, $\alpha$ is called an exceptional value of $f(z)$ at $z_0$ for simplicity. In this paragraph, we assume that $\Gamma - \Gamma'$ and $\Gamma - \overline{(\Gamma - E)}$ consist of only essential singularities of $f(z)$[1]. Accordingly, the set $\Omega = C_D(f, z_0) - C_{\Gamma - E}(f, z_0)$ is an open set. Hervé's method is to investigate properties of the set $E$ and the cluster sets $C\ (f, \zeta)$ (e. g., in the case where $\Delta$ is a simple curve) which can be deduced from the hypothesis that a connected component of $\Omega = C_D(f, z_0) - C_{\Gamma - E}(f, z_0)$ contains at least three exceptional values.

We state Hervé's theorems without proofs[2].

**Theorem 3.** *If $z_0 \in E$, if a connected component of $\Omega = C_D(f, z_0) - C_{\Gamma - E}(f, z_0)$ contains three exceptional values $w$, $w'$, $w^*$ of $w = f(z)$ at $z_0$ and if $U$ is a simply connected domain containing these three values such that $\bar{U} \subset \Omega$, then it is possible to find a perfect subset $P$ of $E$, containing $z_0$, with the following properties:*

(i) *if $\zeta \in P$, then $U \subset C_D(f, \zeta) - C_{\Gamma - E}(f, \zeta)$;*

(ii) *if $\zeta \in \Gamma - P$ and $|\zeta - z_0|$ is sufficiently small, then $U \cap C_D(f, \zeta) = \emptyset$*[3]. (HERVÉ [3]).

**Theorem 4.** *Under the same hypothesis as in Theorem 3, there exist a positive number $\varrho$ and, in the perfect set $P$ defined in Theorem 3, a subset $p_1$ of the first category on $P$, which possess the following property: If $\xi \in P$ and if $\Delta$ is a simple curve described in $D$, terminating at $\xi$, such that the diameter of $C_\Delta(f, \xi)$ is less than $\varrho$, then $\xi \in p_1$. In particular, all the points of $P$ where $w = f(z)$ admits asymptotic values belong to $p_1$* (HERVÉ [3]).

Using a lemma, closely related to the Lemma in Paragraph 2, HERVÉ has proved a more general theorem. For this purpose, we define a subset $p(m, X)$ of $P$ in the following way: $\xi \in p(m, X)$ if there exists a continuum $\Xi$ such that (i) $\xi \in \Xi$; (ii) the diameter of $\Xi > 1/m$ ($m$: entire); (iii) $z \in \Xi - \Gamma$ implies $z \in D$ and $f(z) \in X$.

**Theorem 5.** *Under the same assumption as above, the set $p(m, X)$ is nowhere dense on $P$ for every closed set $X$ such that a connected component of $U - X$ contains two exceptional values $w'$ and $w^*$ of $f(z)$ at $z_0$* (HERVÉ [3]).

**Theorem 6.** *If $\alpha \in C_D(f, z_0) - C_{\Gamma - E}(f, z_0)$, if $\alpha = \lim_{n \to \infty} \alpha_n$ and if $f(z) \neq \alpha_n$ for $z \in D$ and $|z - z_0| < r$, where $r$ is a fixed positive number, then all the points of $E$ where $w = f(z)$ admits $\alpha$ as an asymptotic value form a set which contains a perfect set* (HERVÉ [3]).

---

[1] $\Gamma'$ denotes the derived set of $\Gamma$.

[2] For the proofs, cf. HERVÉ [3].

[3] From this theorem follows that if the intersection of $E$ with a neighborhood of $z_0$ is countable, then every component of $\Omega$ contains at most two exceptional values of $f(z)$ at $z_0$ (cf. Theorem 1, § 5).

# III. Functions meromorphic in the unit circle

## § 1. Functions of class $(U)$ in Seidel's sense

**1.** Let $w = f(z)$ be a bounded regular function in the unit circle $|z| < 1$. Then, by Fatou's theorem, the radial limit $\lim_{r\to 1} f(re^{i\theta}) = f(e^{i\theta})$ exists for all values of $\theta$ in $0 \leqq \theta \leqq 2\pi$ except perhaps for a set of values of $\theta$ of measure zero. If the modulus of the boundary function $f(e^{i\theta})$ is equal to unity, i. e. $|f(e^{i\theta})| = 1$, for almost every $\theta$ in $0 \leqq \theta \leqq 2\pi$, then we call $f(z)$ a *function of class* $(U)$ in the sense of SEIDEL[1].

Independently, SEIDEL [2] and FROSTMAN [1] have made important contributions to the theory of functions of class $(U)$. Their methods are based on the integral representation for a function $f(z)$ of class $(U)$. Suppose, for simplicity, that $f(0) \neq 0$ and denote its zero-points (if they exist) by $a_1, a_2, \ldots, a_k, \ldots$. Obviously $\sum_k (1 - |a_k|) < \infty$. We form the Blaschke produkt

$$B(z) = \prod_k \frac{\bar{a}_k}{|a_k|} \frac{a_k - z}{1 - \bar{a}_k z}$$

which is a function of class $(U)$[2]. If we define $g(z)$ by the relation $f(z) = B(z)\, g(z)$, then $g(z)$ is also a function of class $(U)$ and does not vanish in $|z| < 1$. Consider now the function $h(z) = -\log g(z)$, selecting a definite branch of the logarithm. Then, $h(z)$ is single-valued and regular in $|z| < 1$. Furthermore, $\Re\, h(z) \geqq 0$ in $|z| < 1$. By Herglotz's theorem[3], we have

$$h(z) = \frac{1}{2\pi} \int_0^{2\pi} \frac{e^{i\theta} + z}{e^{i\theta} - z}\, d\mu(\theta) + i\gamma\,,$$

where $\mu(\theta)$ is a monotonically non-decreasing function of $\theta$ in the interval $0 \leqq \theta \leqq 2\pi$ and $\gamma$ is a real constant. It is well-known that $v(z) = \Re h(z) = -\log|g(z)|$ has an angular limit $\mu'(\theta)$ for all values of $\theta$ in $0 \leqq \theta \leqq 2\pi$ for which $\mu(\theta)$ has a derivative. Since $g(z)$ belongs to class $(U)$, the radial limit of $\Re h(z)$ at $e^{i\theta}$, and therefore $\mu'(\theta)$, is equal to zero almost everywhere in $0 \leqq \theta \leqq 2\pi$.

**Theorem 1.** *Let $w = f(z)$ be a function of class $(U)$. Then*

$$f(z) = e^{-i\gamma} B(z) \exp\left[-\frac{1}{2\pi} \int_0^{2\pi} \frac{e^{i\theta} + z}{e^{i\theta} - z}\, d\mu(\theta)\right], \tag{1}$$

*where $B(z)$ is the Blaschke product extended over the zero-points of $f(z)$, $\mu(\theta)$ is a monotone non-decreasing function of $\theta$ in $0 \leqq \theta \leqq 2\pi$ whose*

[1] NEVANLINNA [1] was the first to point out the interest which lies in the class $(U)$. Cf. HÖSSJER and FROSTMAN [1], SEIDEL [2], FROSTMAN [1], NOSHIRO [3, 11], CALDERÓN, GONZÁLES DOMÍNGUES and ZYGMUND [1], LOHWATER [2, 3, 7], OHTSUKA [9].

[2] F. RIESZ [1].

[3] HERGLOTZ [1].

*derivative* $\mu'(\theta)$ *is equal to zero almost everywhere in* $0 \leqq \theta \leqq 2\pi$ *and* $\gamma$ *is a real constant* (SEIDEL [2], FROSTMAN [1]).

Suppose that $g(z)$ is not a constant. Then, the function $\mu(\theta)$ is not identically constant. Since $\mu(\theta)$ is monotonically non-decreasing and $\mu'(\theta) = 0$ almost everywhere in $0 \leqq \theta \leqq 2\pi$, it admits the following representation: $\mu(\theta) = \eta(\theta) + \nu(\theta)$. Here $\eta(\theta)$ and $\nu(\theta)$ are non-decreasing; $\eta(\theta)$ is continuous and $\eta'(\theta) = 0$ almost everywhere in $0 \leqq \theta \leqq 2\pi$ and $\nu(\theta)$ a step-function. If $\eta(\theta)$ is not constant, the symmetric derivative of $\eta(\theta)$ is $+\infty$ at a non-enumerable set of points[1].

Using these facts, we prove that there exists a point $e^{i\varphi}$ for which the radial limit of $g(z)$ is equal to zero.

If $|\theta - \varphi| \leqq 1 - r = \delta$, then, by an elementary calculation,

$$\Re \frac{e^{i\theta} + z}{e^{i\theta} - z} = \frac{1 - r^2}{1 - 2r\cos(\theta - \varphi) + r^2} > \frac{1}{1 - r} = \frac{1}{\delta}, \quad (z = re^{i\varphi}).$$

Accordingly,

$$v(z) = \Re g(z) \geqq \frac{1}{2\pi} \int_{\varphi-\delta}^{\varphi+\delta} \frac{1 - r^2}{1 - 2r\cos(\theta - \varphi) + r^2}\, d\mu(\theta)$$

$$> \frac{1}{2\pi\delta} \int_{\varphi-\delta}^{\varphi+\delta} d\mu(\theta).$$

If $\varphi$ is a point of discontinuity of $\mu(\theta)$, then we have for all $\delta$

$$\int_{\varphi-\delta}^{\varphi+\delta} d\mu(\theta) \geqq m > 0,$$

for a fixed positive number $m$; hence $\lim_{r\to 1} v(re^{i\varphi}) = +\infty$. Suppose now that $\mu(\theta) \equiv \eta(\theta)$. Then, there exists a point $\varphi$ for which the symmetric derivative of $\mu(\theta)$ is equal to $+\infty$. Consequently,

$$\lim_{r\to 1} v(re^{i\varphi}) \geqq \lim_{\delta\to 0} \frac{1}{2\pi\delta} \int_{\varphi-\delta}^{\varphi+\delta} d\mu(\theta) = \frac{1}{2\pi} \lim_{\delta\to 0} \frac{\mu(\varphi+\delta) - \mu(\varphi-\delta)}{\delta} = +\infty.$$

Thus

$$\lim_{r\to 1} |g(re^{i\varphi})| = \lim_{r\to 1} e^{-v(re^{i\varphi})} = 0.$$

**Theorem 2.** *If* $f(z)$ *is a non-constant function of class* $(U)$ *and if* $f(z)$ *is not a Blaschke product, then* $f(z)$ *admits* 0 *as its radial limit* (FROSTMAN [1]).

Remark. It should be noted that there exists a Blaschke product which admits 0 as its radial limit. FROSTMAN [1] has constructed the

[1] SCHLESINGER and PLESSNER [1], § 43.

following Blaschke product:

$$B(z) = \prod_{k=1}^{\infty} \frac{\left(1-\frac{1}{k^2}\right)-z}{1-\left(1-\frac{1}{k^2}\right)z}$$

which has the radial limit 0 at $z = 1$.

As an immediate consequence, we have

**Theorem 3.** *Let $f(z)$ be a non-constant function of class $(U)$. If $f(z) \neq \alpha$ $(|\alpha| < 1)$ in $|z| < 1$, then there exists at least one radius $\theta = \varphi$ such that* $\lim_{r\to 1} f(re^{i\varphi}) = \alpha$ (SEIDEL [2])[1].

2. Now we shall study functions of class $(U)$ by a method[2], different from those of SEIDEL and FROSTMAN, which is rather topological and does not depend on the Poisson-Stieltjes-Herglotz integral representation of bounded regular functions in the unit circle. For that purpose, we first state an extension of Löwner's lemma[3].

**Lemma 1.** *Let $z = z(\zeta)$ be a function regular and bounded: $|z(\zeta)| < 1$ in the unit circle $|\zeta| < 1$ such that $z(0) = 0$. Let $E_\zeta$ be a set on $|\zeta| = 1$ such that for every $e^{i\vartheta} \in E_\zeta$ the radial limit $z(e^{i\vartheta})$ is of modulus unity. Denote by $E_z$ the set of values $z(e^{i\vartheta})$ for all $e^{i\vartheta} \in E_\zeta$. Then $m_* E_\zeta \leqq m^* E_z$, where $m_* E_\zeta$ and $m^* E_z$ denote the interior measure of $E_\zeta$ and the exterior measure of $E_z$ respectively.*

*Proof.* For any positive number $\varepsilon$, there exists an open set $G$ on $|z| = 1$ such that $E_z \subset G$, $mG < m^* E_z + \varepsilon$. Let $E'_\zeta$ be a closed subset of $E_\zeta$. Consider two harmonic measures $\Omega(z)$ and $\omega(\zeta)$ of $G$ and $E_\zeta$ with respect to the unit disc respectively. Obviously $V(\zeta) \equiv \Omega(z(\zeta)) - \omega(\zeta)$ is harmonic and bounded in $|\zeta| < 1$ such that the radial limit $V(e^{i\vartheta}) \geqq 0$ for almost every $e^{i\vartheta}$. Consequently, $V(0) = \Omega(z(0)) - \omega(0) = \Omega(0) - \omega(0) \geqq 0$; whence follows $mE'_\zeta \leqq mG < m^* E_z + \varepsilon$. Thus, we have $m_* E_\zeta \leqq m^* E_z$.

**Corollary.** *If we omit the assumption $z(0) = 0$ in the above lemma, we can assert that $m^* E_z > 0$ provided that $m_* E_\zeta > 0$.*

We shall give another proof of the corollary. Denote by $D$ and $\Gamma$ the unit disc $|z| < 1$ and the circumference $|z| = 1$ respectively.

**Lemma 2.** *Let $E$ be a closed set of linear measure zero on $\Gamma$. Then, there exists a function $u(z)$ (an analogue of Evans-Selberg's potential) such that $u(z)$ is positive and harmonic on $D \cup (\Gamma - E)$ and the boundary value of $u(z)$ at every point of $E$ is $+\infty$* (F. and M. RIESZ [1]).

*Proof.* To construct such a function $u(z)$, let $\Gamma - E$ consist of a sequence of open arcs $(e^{i\theta};\, a_n < \theta < b_n)$ $(n = 1, 2, \ldots)$. We define a

---

[1] Remark that $\dfrac{f(z)-\alpha}{1-\bar{\alpha} f(z)}$ is a function of class $(U)$.

[2] NOSHIRO [11]. Cf. also LOHWATER [2].

[3] HÖSSJER [1], KAWAKAMI [1], KAMETANI-UGAERI [1], LOHWATER-SEIDEL [1], OHTSUKA [2], TSUJI [17].

boundary function $V(\theta)$ in the following way: If $a_n < \theta < b_n$, $V(\theta) = c_n \psi_n(\theta) + d_n$, where $\psi_n(\theta) = [(b_n - \theta)(\theta - a_n)]^{-\frac{1}{2}}$, and if $e^{i\theta} \in E$, $V(\theta) = +\infty$, where two sequences of positive numbers $\{c_n\}$, $\{d_n\}$ are chosen so that $\sum_{n=1}^{\infty} c_n \int_{a_n}^{b_n} \psi_n(\theta)\, d\theta < +\infty$, $\sum_{n=1}^{\infty} d_n(b_n - a_n) < +\infty$ and $d_n$ increases monotonically to infinity. Then, the Poisson integral $u(z)$ with the boundary function $V(\theta)$ is a required harmonic function.

Obviously the function

$$w = \chi(z) = e^{-(u(z) + iv(z))},$$

which is an analogue of Evans-Selberg's function, is regular and bounded: $|\chi(z)| < 1$ on $D \cup (\Gamma - E)$ and possesses the boundary value 0 at every point of $E$[1]. The following is remarkable: Denote by $\Gamma_\lambda$ the level curve $u(z) = \text{const.} = \lambda$ $(\inf u < \lambda < +\infty)$. Suppose that $\int_{\Gamma_\lambda} dv(z)$ is bounded. Then $E$ must be of capacity zero (Kuramochi and Kuroda [1]).

Using the function $w = \chi(z)$, we can prove the preceding corollary. If $m_* E_\zeta > 0$, then there exists a closed subset $E'_\zeta$ of positive measure of $E_\zeta$ such that $z(e^{i\vartheta})$ is continuous on $E'_\zeta$ (by Egoroff's theorem). The image $E'_z$ of $E'_\zeta$ is a closed subset of $E_z$. If $E'_z$ is of linear measure zero and if $w = \chi(z)$ is the function associated with $E'_z$, then the composed function $w = \chi(z(\zeta))$ is a non-constant bounded regular function with radial limit 0 at every point of the set $E'_\zeta$ of positive measure. This contradicts the well-known Riesz theorem. Consequently $E'_z$ is of positive measure and hence $m^* E_z > 0$.

Let $f(z)$ be a function of class $(U)$. Then it is easily proved that if $f(z)$ has a singularity at $z_0 = e^{i\theta_0}$, the cluster set $C_D(f, z_0)$ of $f(z)$ at $z_0$ is the closed disc $|w| \leqq 1$[2].

**Theorem 4.** *Let $w = f(z)$ be a non-constant function of class $(U)$ and $(c)$ be any circular disc: $|w - \alpha| < \varrho$ lying inside $|w| < 1$ whose periphery may be tangent to the circumference $|w| = 1$. Denote by $\Delta$ any connected component of the inverse image of $(c)$ under $w = f(z)$ and by $z = z(\zeta)$ a function*

[1] In the case where $E$ is a compact set of capacity zero, some important related theorems have been obtained by Beurling [2], Mori [1, 9], Tsuji [8], Kaplan [1], Pfluger [3, 4]. Mori [1] has proved that if $E$ is a compact set of capacity zero on $\Gamma$, then there exists a univalent regular function $w = \chi(z)$ with the following properties: (1) $w = \chi(z)$ maps $D$ conformally onto a domain $\mathfrak{D}$, starshaped with respect to $w = 0$, which is the whole $w$-plane with an enumerable infinity of infinite radial slits, which cluster to $w = \infty$ only. (2) Every point of $E$ corresponds to an accessible boundary point of $\mathfrak{D}$ lying on $w = \infty$. (3) $\chi(0) = 0$. The following special case of Beurling's theorem is also important: Let $E$ be a compact set on $\Gamma$. Let $w = \chi(z)$ be a univalent regular function in $D$ such that $\lim_{z \to z_0} \chi(z) = \infty$ for every point $z_0$ in $E$. Then $E$ has capacity zero (Kaplan [1], p. 21).

[2] Seidel [2].

*which maps* $|\zeta| < 1$ *onto the simply connected domain* $\Delta$ *in a one-to-one conformal manner. Then, the function*

$$W = F(\zeta) \equiv \frac{1}{\varrho}\,[f(z(\zeta)) - \alpha]$$

*is also a function of class* $(U)$. *(In other words, if* $f(z)$ *is a function of class* $(U)$, *then* $f(z)$ *is also locally of class* $(U)$.*)* (NOSHIRO [6].)

*Proof.* If the closure of $\Delta$ lies in $D$: $|z| < 1$, the assertion is evidently true. Hence, we have only to treat the case where $\Delta$ has at least one boundary point on $\Gamma$: $|z| = 1$. Consider $z = z(\zeta)$ and $W = F(\zeta)$, which are regular and bounded in $|\zeta| < 1$, and denote by $E_\zeta$ the set of $e^{i\theta}$ for which both radial limits $z(e^{i\theta})$, $F(e^{i\theta})$ exist and furthermore $|F(e^{i\theta})| < 1$. $E_\zeta$ is obviously a Borel set and the set $E_z$ of the radial limits $z(e^{i\theta})$ for all $e^{i\theta} \in E_\zeta$ lies on $|z| = 1$. If $mE_\zeta > 0$, then, by the preceding corollary, $m^* E_z > 0$. However, for every point $e^{i\theta} \in E_z$, the radial limit $f(e^{i\theta})$ lies in the interior of $(c)$: $|w - \alpha| < \varrho$. This leads to a contradiction, since $f(z)$ is a function of class $(U)$. Accordingly, $E_\zeta$ is of linear measure zero; i. e., $F(\zeta)$ is also a function of class $(U)$.

As an application of Theorem 4, we state

**Theorem 5.** (an extension of Iversen's theorem). *Let* $w = f(z)$ *be a function of class* $(U)$ *and* $z = \varphi(w)$ *be its inverse function defined in the unit circle* $|w| < 1$. *Then, for any disc* $(c)$: $|w - \alpha| < \varrho$ *lying inside* $|w| < 1$ *and for any element* $e(w, w_0)$ *of* $z = \varphi(w)$ *with center* $w_0$ *lying in* $(c)$, *it is possible to find a suitable path joining* $w = w_0$ *and* $w = \alpha$ *inside* $(c)$ *along which the element* $e(w, w_0)$ *can be continued analytically except perhaps at* $w = \alpha$. *(In other words,* $z = \varphi(w)$ *has Iversen's property inside* $|w| < 1$ *in the sense of* STOÏLOW [9].*)* (NOSHIRO [6].)[1]

*Proof.* Let $z = \varphi_\varrho(w)$ be the branch of $z = \varphi(w)$ defined by the element $e(w, w_0)$ inside $(c)$: $|w - \alpha| < \varrho$ and $\Delta$ be the set of values of $z = \varphi_\varrho(w)$. Then two cases will be considered: Either $f(z)$ assumes the value $\alpha$ at least once or $f(z)$ omits $\alpha$ in $\Delta$. In the former case, obviously there exists a path $\gamma_w$, connecting $w = w_0$ and $w = \alpha$ inside $(c)$, along which the continuation of $e(w, w_0)$ is possible including $w = \alpha$. In the

[1] We can prove indirectly that there exists a function $w = f(z)$ of class $(U)$, whose inverse function $z = \varphi(w)$ has no Gross' property in $|w| < 1$. A Riemann surface $F$ is said to belong to $O_{HB}$ provided that there exists no non-constant single-valued bounded harmonic function on $F$. It is known that there exists a covering surface $\Phi$ of the $w$-plane, belonging to $O_{HB}$, which has no Gross' property (cf. for example, KURAMOCHI [3]). Suppose that $P_0$ is a regular point of $\Phi$ above $w = w_0$ and that the set of arguments $\psi = \arg(w - w_0)$ of singular segments starting from $P_0$ above the disc $|w - w_0| < R$ is of positive measure. Consider a connected piece $\Phi_R$ of $\Phi$ containing $P_0$ above $|w - w_0| < R$ and map the unit disc $|z| < 1$ conformally onto the universal covering surface $\tilde{\Phi}_R$ of $\Phi_R$; we denote by $w = g(z)$ the mapping function. Then $W = f(z) = R^{-1}(g(z) - w_0)$ is a function of class $(U)$ and its inverse $z = \varphi(W)$ has no Gross' property.

latter case, denote by $z = z(\zeta)$ the univalent function mapping $|\zeta| < 1$ conformally onto $\Delta$. Then, by Theorem 4, $F(\zeta) \equiv \frac{1}{\varrho}[f(z(\zeta)) - \alpha]$ is a function of class ($U$). Accordingly, the value $\alpha$ is a cluster value of $w = f(z(\zeta))$ at a certain point on $|\zeta| = 1$; hence $e(w, w_0)$ can be continued at a point, say $w_1$, inside $(c_1)$: $|w - \alpha| < \varrho/2$ by a suitable path lying entirely inside $(c)$: $|w - \alpha| < \varrho$. Next, consider the branch of $z = \varphi_\varrho(w)$ defined by the element $e(w, w_1)$ just obtained at $w = w_1$ within the circle $(c_1)$. The same argument shows that $e(w, w_1)$ can be continued at a point $w = w_2$ inside $(c_2)$: $|w - \alpha| < \varrho/2^2$ by a suitable path inside $(c_1)$: $|w - \alpha| < \varrho/2$. Thus, repeating the same arguments indefinitely, we can find a path $\gamma_w$, lying in $(c)$, along which the continuation of $e(w, w_0)$ is possible except the end-point $w = \alpha$ and defines a transcendental singularity at $w = \alpha$.

As an immediate consequence of Theorem 5, we have

**Theorem 6.** *Let $w = f(z)$ be a non-constant function of class* ($U$). *Then,*

(i) *the set of all the radial limits $f(e^{i\theta})$ contains the circumference* $|w| = 1$;

(ii) *if $w = f(z)$ omits a value $\alpha$* ($|\alpha| < 1$) *in* $|z| < 1$, *then there exists at least one point $e^{i\theta}$ such that* $\lim_{r \to 1} f(re^{i\theta}) = \alpha$;

(iii) *if $w = f(z)$ has at least one singularity on* $|z| = 1$ *and if $f(z)$ assumes $\alpha$ only finitely often in* $|z| < 1$, *the same assertion as in* (ii) *holds* (SEIDEL [2]).

Remark. Theorem 3 is contained in Theorem 6. OHTSUKA [6] has constructed a function of class ($U$) which admits every value of modulus $< 1$ as a radial limit.

As another application of Theorem 4, we give a proof for Seidel-Frostman's theorem.

**Theorem 7.** *Let $w = f(z)$ be a function of class* ($U$). *Suppose that $f(z)$ has a singularity at $z_0 = e^{i\theta_0}$. Then, every value of $D_w$: $|w| < 1$ is assumed by $f(z)$ infinitely often in any neighborhood of $z_0$ except perhaps for a set of values of capacity zero; i. e., $D_w - R_D(f, z_0)$ is at most of capacity zero* (SEIDEL [2], FROSTMAN [1]).

*Proof.* Suppose that $D_w - R_D(f, z_0)$ is of positive capacity. Then there is a positive number $r$ such that the set $E_w$ of exceptional values of $w = f(z)$ in $D \cap U(z_0, r)$ is of positive capacity, where $U(z_0, r)$ denotes $|z - z_0| < r$. Let $\alpha \in E_w$ be a point such that for every $\varrho > 0$, the part of $E_w$ inside $U(\alpha, \varrho)$: $|w - \alpha| < \varrho$ is of positive capacity. Choose a positive number $r_0 (< r)$ so that $|f(z) - \alpha| > \varrho_0$ on $|z - z_0| = r_0$ in $D$ where $\varrho_0$ is a suitable positive number. We may assume that $(c_0)$: $|w - \alpha| < \varrho_0$ lies in $D_w$. Since $\alpha \in C_D(f, z_0)$, there exists a connected component $\Delta$ of the inverse image of $(c_0)$ inside $D \cap U(z_0, r_0)$. We denote by $E_w^0$ a closed subset of $E_w$, lying completely inside $(c_0)$, of positive capacity. Given the boundary value 0 at every point of $c_0$: $|w - w_0| = \varrho_0$ and 1 at every point of $E_w^0$, there exists a non-constant bounded harmonic function $u^*(w)$ in the domain bounded by $c_0$ and $E_w^0$ as the solution for the Dirichlet problem. Composing $u^*(w)$ with $w = f(z(\zeta))$ introduced in Theorem 4, we have a non-constant bounded harmonic function $V(\zeta) \equiv u^*(f(z(\zeta)))$ in $|\zeta| < 1$ which has the radial limit 0 for almost every point $\zeta = e^{i\theta}$. This is a contradiction.

**3. Generalizations.** Seidel-Frostman's results on functions of class $(U)$ have been generalized by OHTSUKA [1, 9], LOHWATER [2, 4, 7], LEHTO [3, 7], STORVICK [1, 2] and others in various ways.

To discuss systematically such generalizations, we first state an important result of BAGEMIHL [1] on cluster sets of arbitrary complex-valued functions.

**Lemma 3.** *Let $D$ and $\Gamma$ be the unit disc $|z| < 1$ and the circumference $|z| = 1$ respectively. If $S$ is an arbitrary subset of $D$, then there exists a subset $\Gamma^*$ of $\Gamma$, with $\Gamma - \Gamma^*$ at most countable, such that, for every $e^{i\theta} \in \Gamma^*$, if $\Lambda_1$ and $\Lambda_2$ are two simple curves in $D$ terminating at $e^{i\theta}$, either $\Lambda_1$ and $\Lambda_2$ both intersect $S$ or both intersect $\mathscr{C}S = D - S$* (BAGEMIHL [1]).

*Proof*[1]. For every pair of rational numbers $\alpha$, $\varrho$ satisfying

$$0 < \alpha < \frac{\pi}{2}, \quad 0 < \varrho < 1, \tag{1}$$

we define a set $E_{\alpha,\varrho}$ on $\Gamma$ as follows: $e^{i\theta}$ belongs to $E_{\alpha,\varrho}$ if there exist two simple curves $\Lambda$, $\Lambda'$ ending at $e^{i\theta}$, with respective end points $\zeta$, $\zeta'$ in $D$, such that

$$|\zeta| = |\zeta'| = \varrho, \quad \theta - \frac{\pi}{4} < \arg\zeta' < \theta + \frac{\pi}{4} - \alpha < \arg\zeta < \theta + \frac{\pi}{4}; \tag{2}$$

for every $z \in \Lambda \cup \Lambda'$ different from $\zeta$ and $\zeta'$,

$$\varrho < |z| < 1, \quad \theta - \frac{\pi}{4} < \arg z < \theta + \frac{\pi}{4} \tag{3}$$

and

$$\Lambda \subset S, \quad \Lambda' \subset \mathscr{C}S. \tag{4}$$

Define the set $E'_{\alpha,\varrho}$ similarly, replacing (4) by

$$\Lambda \subset \mathscr{C}S, \quad \Lambda' \subset S. \tag{5}$$

Obviously, $E_{\alpha,\varrho}$ and $E'_{\alpha,\varrho}$ are isolated sets and are therefore at most countable. Let $\Gamma^*$ be the set of all $e^{i\theta}$ which do not belong to any $E_{\alpha,\varrho}$ or $E'_{\alpha,\varrho}$, with $\alpha$ and $\varrho$ rational numbers satisfying relations (1). Since there are only countably many such sets $E_{\alpha,\varrho}$, $E'_{\alpha,\varrho}$ and each of these sets is at most countable, $\Gamma - \Gamma^*$ is at most countable.

Now, suppose that $e^{i\theta} \in \Gamma^*$ and that $\Lambda_1$ and $\Lambda_2$ are two simple curves in $D$ terminating at $e^{i\theta}$. If $\Lambda_1$ and $\Lambda_2$ have at least one point in common,

[1] This proof is due to BAGEMIHL [1].

then the assertion of Lemma 3 is evidently true. Assume, then, that $\Lambda_1$ and $\Lambda_2$ do not intersect. Starting at $e^{i\theta}$, proceed along $\Lambda_1$ until the first point of $g_{\theta-\frac{\pi}{4}} \cup g_{\theta+\frac{\pi}{4}}$ is reached[1], and denote by $\Lambda^*_1$the curve ending at $e^{i\theta}$ thus described; if $\Lambda_1$ intersects neither of radii $g_{\theta\pm\frac{\pi}{4}}$, put $\Lambda_1^* = \Lambda_1$. Define $\Lambda_2^* \subset \Lambda_2$ analogously. Select a rational number $\varrho$, $0 < \varrho < 1$, so large that the circle $|z| = \varrho$ intersects both $\Lambda_1^*$ and $\Lambda_2^*$ and contains in its interior the end points in $D$ of these two curves. Starting at $e^{i\theta}$, proceed along $\Lambda_1^*$ until the first point, say $\zeta$, of $|z| = \varrho$ is reached, and denote by $\Lambda$ the arc thus described; proceeding along $\Lambda_2^*$ from $e^{i\theta}$, define the point $\zeta'$ and the arc $\Lambda'$ terminating at $e^{i\theta}$. Then $\zeta \neq \zeta'$, and we may assume that, as viewed from the origin, $\zeta$ lies to the left of $\zeta'$ on $|z| = \varrho$. On the minor arc of $|z| = \varrho$ determined by the points $\varrho\, e^{i(\theta\pm\pi/4)}$, select a point $\beta$ between $\zeta$ and $\zeta'$ such that $\alpha = \theta + \frac{\pi}{4} - \arg\beta$ is a rational number between 0 and $\pi/2$. We have relations (1) and (2), and relation (3) holds for every $z \in \Lambda \cup \Lambda'$, except for $\zeta$ and $\zeta'$. Neither condition (4) nor condition (5) is satisfied; for otherwise, we would have $e^{i\theta} \in E_{\alpha,\varrho} \cup E'_{\alpha,\varrho}$, which contradicts the assumption that $e^{i\theta} \in \Gamma^*$. Hence, either $\Lambda$ and $\Lambda'$ both intersect $S$ or both intersect $\mathscr{C} S$, which implies that the same is true of $\Lambda_1$ and $\Lambda_2$.

Let $w = f(z)$ be an arbitrary complex-valued function defined in $D$: $|z| < 1$. If $\Lambda$ is a simple curve in $D$ terminating at $e^{i\theta}$, then we define the cluster set $C_\Lambda(f, e^{i\theta})$ of $f(z)$ at $e^{i\theta}$ along $\Lambda$ in the following way: $\alpha \in C_\Lambda(f, e^{i\theta})$, if there exists a sequence of points $\{z_n\}$ on $\Lambda$ such that $z_n \to e^{i\theta}$ and $f(z_n) \to \alpha$ as $n \to \infty$.

If there are two simple curves $\Lambda_1$ and $\Lambda_2$ in $D$ terminating at $e^{i\theta}$ such that

$$C_{\Lambda_1}(f, e^{i\theta}) \cap C_{\Lambda_2}(f, e^{i\theta}) = \emptyset$$

then we call $z = e^{i\theta}$ an *ambiguous point* of $f(z)$ (Bagemihl and Seidel [6]).

**Theorem 8.** *If $w = f(z)$ is an arbitrary complex-valued function defined in $D$: $|z| < 1$, then there exists a set $\Gamma^*$ on $\Gamma$: $|z| = 1$ with $\Gamma - \Gamma^*$ at most countable, such that, for every $e^{i\theta} \in \Gamma^*$ and every pair of simple curves $\Lambda_1$, $\Lambda_2$ lying in $D$ and terminating at $e^{i\theta}$,*

$$C_{\Lambda_1}(f, e^{i\theta}) \cap C_{\Lambda_2}(f, e^{i\theta}) \neq \emptyset\,.$$

*Hence, the set of ambiguous points of $f(z)$ on $\Gamma$: $|z| = 1$ is at most countable* (Bagemihl [1]).

*Proof*[2]. Consider a basis consisting of a countable number of neighborhoods $\{V_n\}$ on the Riemann $w$-sphere $\Sigma$. Then, countably many open sets possess the property being expressible as the union of finitely many

[1] We denote by $g_\varphi$ the radius: $\arg z = \varphi$, $0 \leq |z| < 1$.

[2] This proof is due to Bagemihl [1].

$V_n$; let these open sets be $G_1, G_2, \ldots, G_n, \ldots$. For every natural number $n$, set

$$S_n = \{z : z \in D, f(z) \in G_n\}.$$

According to Lemma 3, there exists a set $\Gamma_n^* \subset \Gamma$, with $\Gamma - \Gamma_n^*$ at most countable, such that, for every $e^{i\theta} \in \Gamma_n^*$, if $\Lambda_1'$ and $\Lambda_2'$ are simple curves lying in $D$ and terminating at $e^{i\theta}$, then either $\Lambda_1'$ and $\Lambda_2'$ both intersect $S_n$ or both intersect $\mathscr{C} S_n = D - S_n$. If we put $\Gamma^* = \bigcap_{n=1}^{\infty} \Gamma_n^*$, then $\Gamma - \Gamma^*$ is at most countable. Now, suppose that $e^{i\theta} \in \Gamma^*$ and that $\Lambda_1$ and $\Lambda_2$ are simple curves in $D$ terminating at $e^{i\theta}$. Assume that $C_{\Lambda_1}(f, e^{i\theta}) \cap C_{\Lambda_2}(f, e^{i\theta}) = \emptyset$. Then there exists a natural number $n$ such that

$$C_{\Lambda_1}(f, e^{i\theta}) \subset G_n, \quad C_{\Lambda_2}(f, e^{i\theta}) \subset \Sigma - \overline{G}_n.$$

If we consider suitable last parts $\Lambda_1^*$, $\Lambda_2^*$ of $\Lambda_1$, $\Lambda_2$, these images of $\Lambda_1^*$ and $\Lambda_2^*$ by $w = f(z)$ lie completely in $G_n$ and $\Sigma - G_n$ respectively. Whence follows that $\Lambda_1^* \subset S_n$ and $\Lambda_2^* \subset \mathscr{C} S_n$. This is a contradiction.

Remark. Given a countable set of points $\{\zeta_n\}$ on $|z| = 1$, there exists a function $f(z)$, regular and of bounded type in $|z| < 1$, for which every point $\zeta_n$ is an ambiguous point with the following property: there exist simple curves $\Lambda_1^{(n)}$, $\Lambda_2^{(n)}$ in $|z| < 1$ terminating at $\zeta_n$ for which $f(z)$ admits different asymptotic values (Bagemihl and Seidel [6]).

Now let $w = f(z)$ be a meromorphic function in the unit disc $D$: $|z| < 1$. Let $z_0 = e^{i\theta_0}$ be a fixed point on $\Gamma$: $|z| = 1$ and $A$ an open arc of $\Gamma$ containing $z_0$. We suppose that $E$ is a set of measure zero containing $z_0$ and contained in $A$. We associate with every $e^{i\theta} \in A - E = \mathscr{C} E$ an arbitrary curve $\Lambda_\theta$ in $D$ terminating at $z = e^{i\theta}$ and the cluster set $C_{\Lambda_\theta}(f, e^{i\theta})$ of $f(z)$ at $e^{i\theta}$ along $\Lambda_\theta$. Clearly $C_{\Lambda_\theta}(f, e^{i\theta})$ is either a continuum or a single point. We define a new *boundary cluster set* $C^*_{\Gamma-E}(f, z_0)$ of $f(z)$ at $z_0$ as follows: $C^*_{\Gamma-E}(f, z_0) = \bigcap_{r>0} M_r$, where $M_r$ denotes the closure of the union $\bigcup C_{\Lambda_\theta}(f, e^{i\theta})$ for all $e^{i\theta}$ in the intersection of $\mathscr{C} E$ with $|z - z_0| < r$. It is clear that $C^*_{\Gamma-E}(f, z_0) \subset C_\Gamma(f, z_0) \subset C_D(f, z_0)$.

As an analogue of Theorem 4, § 4, II, we have

**Theorem 9.** *$C_D(f, z_0) - C^*_{\Gamma-E}(f, z_0)$ is an open set; i. e., $\mathscr{F} C_D(f, z_0) \subset \subset \mathscr{F} C^*_{\Gamma-E}(f, z_0)$* (Noshiro [11]).

*Proof.* Let $w_0$ be a point belonging to $C_D(f, z_0) - C^*_{\Gamma-E}(f, z_0)$. We first select a positive number $r$ such that $w_0$ is outside $M_r$. Denote by $\varrho'$ the distance of $w_0$ from $M_r$. Next we select two points $z_1 = e^{i\theta_1}$, $z_2 = e^{i\theta_2}$ $(\theta_1 < \theta_0 < \theta_2)$ belonging to $\mathscr{C} E \cap (K)$, where $(K)$ denotes the disc $|z - z_0| < r$. We take a suitable last part $\Lambda'_{\theta_1}$ of $\Lambda_{\theta_1}$ and a suitable last part $\Lambda'_{\theta_2}$ of $\Lambda_{\theta_2}$ lying in $D \cap (K)$. Then it is possible to construct a cross-cut $L$ of $D \cap (K)$ by connecting two end points, lying in $D$, of $\Lambda'_{\theta_1}$ and $\Lambda'_{\theta_2}$ by a rectilinear segment. We may suppose that $f(z) \neq w_0$ on $L$

and that there exists a positive number $\varrho''$ such that $|f(z) - w_0| \geqq \varrho'' > 0$ on $L$. Let $\varrho$ be a positive number less than $\min(\varrho', \varrho'')$. Denote by $D_0$ the domain inclosed by the simple closed curve consisting of $L$ and the arc $(e^{i\theta}, \theta_1 \leqq \theta \leqq \theta_2)$. Since $w_0 \in C_D(f, z_0)$, there exists a sequence of points $z_\mu \in D_0$, $z_\mu \to z_0$, such that $f(z_\mu) \to w_0$. We keep this sequence $z_\mu$ $(\mu = 1, 2, \ldots)$ fixed throughout the proof. The inverse image of $(c)$: $|w - w_0| < \varrho$ in $D_0$ consists of at most a countable number of connected components. We denote the component containing $z_\mu$ by $\Delta_\mu$ which may coincide with other $\Delta_\nu$'s.

We first treat the case in which there is an infinite number of different components $\Delta_\mu$. In this case, we assume that $\Delta_\mu \neq \Delta_\nu$ for $\mu \neq \nu$. If $\Delta_\mu$ is compact in $D_0$, then $(c) = f(\Delta_\mu)$. Suppose that $\Delta_\mu$ is non-compact in $D_0$. It should be noted that $\Delta_\mu$ is not necessarily simply connected. The boundary of $\Delta_\mu$ consists of at most a countable number of analytic curves $\gamma_\mu$ and a closed set on $|z| = 1$. Obviously $|f(z) - w_0| = \varrho$ on $\gamma_\mu$, and $|f(z) - w_0| < \varrho$ in $\Delta_\mu$. We map the universal covering surface $\tilde{\Delta}_\mu$ of $\Delta_\mu$ conformally onto the unit disc $|\zeta| < 1$ in a one-to-one manner. Denote by $z = z(\zeta)$ the mapping function. Considering the two bounded regular functions $z = z(\zeta)$ and $w = w(\zeta) \equiv f(z(\zeta))$, we denote by $E_\zeta$ the set of $\zeta = e^{i\varphi}$ such that both radial limits $z(e^{i\varphi})$, $w(e^{i\varphi})$ exist and $|w(e^{i\varphi}) - - w_0| < \varrho$. Obviously $E_\zeta$ is measurable. If for a point $e^{i\varphi} \in E_\zeta$, $z(e^{i\varphi})$ coincides with some $z = e^{i\theta} \in \mathscr{C}E$, then, $e^{i\theta}$ is an ambiguous point of $f(z)$, since there exists a simple curve $\Lambda$ terminating at $e^{i\theta}$ along which $f(z)$ converges to the asymptotic value $w(e^{i\varphi})$ lying in $(c)$, while $C_{\Delta_\theta}(f, e^{i\theta})$ is situated outside $(c)$. In the other case, $z(e^{i\varphi})$ belongs to $E$ for a point $e^{i\varphi} \in E_\zeta$. Since $E$ is of linear measure zero and $f(z)$ has at most countably many ambiguous points, $E_\zeta$ must be of measure zero by an extension of Löwner's lemma (see, Corollary of Lemma 1, § 1, III). In other words, $(w(\zeta) - w_0)/\varrho$ is a function of class $(U)$. Hence we have $\overline{(c)} = \overline{f(\Delta_\mu)}$. Since there is an infinite number of distinct components $\{\Delta_\mu\}$ converging to $z_0$, we have $\overline{(c)} \subset C_D(f, z_0)$.

Next we consider a sequence of cross-cuts $L_n$ $(n = 1, 2, \ldots)$ converging to $z_0$ and a sequence of positive numbers $\{\varrho_n\}$ tending to zero, where $L_n$ and $\varrho_n$ are chosen as above for each $n$. We denote by $\Delta_\mu^{(n)}$ the component, containing $z_\mu$ $(\mu \geqq N(n))$, of the inverse image of $(c_n)$: $|w - w_0| < \varrho_n$. Suppose that there exists at least one $n$ for which the sequence $\Delta_\mu^{(n)}$ $(\mu \geqq N(n))$ consists of infinitely many domains. We can conclude from the discussion above that $C_D(f, z_0)$ contains $|w - w_0| \leqq \varrho_n$. Thus it remains to consider the case in which, for every $n$, $\Delta_\mu^{(n)}$ consists of a finite number of different domains. Then it is easy to see that there exists a simple curve $\mathscr{L}$ terminating at the point $z_0$ for which $w_0$ is an asymptotic value of $f(z)$. In the case under consideration, we can show that if $\Omega$ denotes the complementary component of $C^*_{\Gamma-E}(f, z_0)$ which contains

$w_0$, then $R_D(f, z_0)$ covers $\Omega$, except for at most a set of capacity zero. Contrary to the assertion, suppose that $\Omega - R_D(f, z_0)$ is of positive capacity. Then we can find a disc $(K_0)$: $|z - z_0| < r_0$ and a set $\Sigma \subset \Omega$ of positive capacity, with the property that $f(z)$ does not belong to $\Sigma$ for any $z \in D \cap (K_0)$. It is clear that there is a point $\alpha \in \Sigma$, different from $w_0$, such that for all positive numbers $\eta$ the intersection $U(\alpha, \eta) \cap \Sigma$ is of positive capacity, $U(\alpha, \eta)$ denoting the $\eta$-neighborhood of $\alpha$. Let $\alpha_1$ be another point belonging to $\Sigma$ and distinct from $\alpha$ and $w_0$. Inside $\Omega$ we draw a simple closed regular analytic curve $Q$ which surrounds $w_0$, $\alpha$ and passes through $\alpha_1$ and whose interior $G$ consists only of interior points of $\Omega$. We put $\Sigma_1 = U(\alpha, \eta_1) \cap \Sigma$, choosing $\eta_1$ so small that $U(\alpha, \eta_1)$ is completely contained in $G$. Now, we select a positive number $r (< r_0)$ such that the set $M_r$ is outside the closure $\overline{G}$ of $G$. Denote by $(K)$ the disc $|z - z_0| < r$. We construct a cross-cut $L$ of $(K) \cap D$ which consists of two last parts $\Lambda'_{\theta_1}$ and $\Lambda'_{\theta_2}$ of $\Lambda_{\theta_1}$ and $\Lambda_{\theta_2}$ $(e^{i\theta_1}, e^{i\theta_2} \in \mathscr{C}E$, $\theta_1 < \theta_0 < \theta_2)$ and a rectilinear segment $s$, using the method stated above. We may suppose that the images of $\Lambda'_{\theta_1}$ and $\Lambda'_{\theta_2}$ have positive distance from $\overline{G}$. In the domain $D_1$ bounded by $L$ and the circular arc $(e^{i\theta}; \theta_1 \leqq \theta \leqq \theta_2)$, we consider the inverse image $f^{-1}(G)$ and denote by $\Delta$ the component of $f^{-1}(G)$ which contains the last part of the asymptotic path $\mathscr{L}$. The boundary of $\Delta$ consists of a finite number of segments $q_i$ $(i = 1, 2, \ldots, m)$ on $s$ and at most a countable number of analytic curves lying in $D_1$ and a bounded closed set on $|z| = 1$. Clearly $\Delta$ is simply connected. We map $\Delta$ by $z = z(\zeta)$ conformally upon the unit disc $|\zeta| < 1$ in a one-to-one manner. The image of $\mathscr{L}$ terminates at a point $\zeta_0$ on $|\zeta| = 1$. We consider a small open arc $A_\zeta$ of $|\zeta| = 1$ containing $\zeta_0$ and disjoint from all $\sigma_i$, where $\sigma_i$ denotes the image of $q_i$. Now, we consider the subset $E_\zeta$ of $A_\zeta$ such that, for $\zeta = e^{i\varphi} \in E_\zeta$, both $z = z(\zeta)$ and $w = w(\zeta) \equiv f(z(\zeta))$ have radial limits $z(e^{i\varphi})$ and $w(e^{i\varphi})$, and such that $w(e^{i\varphi}) \in G$. Suppose that there exists a point $e^{i\varphi} \in E_\zeta$ such that $z(e^{i\varphi})$ coincides with $z = e^{i\theta} \in \mathscr{C}E$. We put $\beta = w(e^{i\varphi})$. Then, by a classical theorem[1] on cluster sets, $\overline{G} \subset C_D(f, e^{i\theta})$, and every value of $G - (\beta)$ is taken infinitely often in any neighborhood of $z = e^{i\theta}$, except for at most two values of $G - (\beta)$. This is a contradiction. Accordingly, $E_\zeta$ must be of linear measure zero, and for almost every $e^{i\varphi} \in A_\zeta$ the radial limit $w(e^{i\varphi})$ of $w(\zeta)$ lies on $Q$. Hence, if we denote by $W = \Phi(w)$ a function mapping $G$ conformally upon the unit disc $|W| < 1$ in a one-to-one manner, the composed function $W = W(\zeta) = \Phi[f(z(\zeta))]$ is a function of generalized class $(U)$. That is, the function $W = W(\zeta)$ is regular and bounded: $|W(\zeta)| < 1$ in $|\zeta| < 1$, and the modulus of the radial limit $W(e^{i\varphi})$ is equal to 1 for almost every $e^{i\varphi}$ in the open arc $A_\zeta$. Furthermore, $W = W(\zeta)$ has a radial limit $W(e^{i\varphi_0})$, $\zeta_0 = e^{i\varphi_0}$, which is equal to $\Phi(w_0)$ lying in $|W| < 1$. By an extension of Seidel-

[1] Cf. Theorem 4, § 4, II and Theorem 7, § 4, II.

Frostman's theorem[1], $W(\zeta)$ takes on every value of $|W| < 1$ infinitely often in any neighborhood of $\zeta_0$, except for at most a set of values of capacity zero. This is a contradiction, since the image $\Phi(\Sigma_1)$ of $\Sigma_1$ is a set of exceptional values of positive capacity.

By arguments similar to those used in the proof of Theorem 9, we can prove the following theorems.

**Theorem 10.** *If $\alpha \in C_D(f, z_0) - C^*_{\Gamma-E}(f, z_0)$ is an exceptional value of $f(z)$ in a neighborhood of $z_0$, then either $\alpha$ is an asymptotic value of $f(z)$ at $z_0$, or there exists a sequence of points $z'_n \in \Gamma$ converging to $z_0$ such that $\alpha$ is an asymptotic value of $f(z)$ at each $z'_n$* (NOSHIRO [11]).

*Proof.* We select a positive number $r$, sufficiently small, such that $f(z) \neq \alpha$ in $D$ for $|z - z_0| < r$ and $\alpha$ lies outside $M_r$. Denote by $\varrho_1$ the distance of $\alpha$ from $M_r$. Next, we select two points $z_1 = e^{i\theta_1}$, $z_2 = e^{i\theta_2}$ $(\theta_1 < \theta_0 < \theta_2)$ belonging to $\mathscr{C}E \cap (K)$, $(K)$ denoting the disc $|z - z_0| < r$. Using the same method as in the proof of Theorem 9, we construct a cross-cut $L$ of $D \cap (K)$ which consists of two last parts of $\Lambda_{\theta_1}$, $\Lambda_{\theta_2}$ and a rectilinear segment $s$. We may suppose that there is a positive number $\varrho_2$ such that $|f(z) - \alpha| \geqq \varrho_2 > 0$ on $L$. Let $\varrho$ be a positive number less than min $(\varrho_1, \varrho_2)$. Denote by $D_0$ the domain surrounded by the simple closed curve consisting of $L$ and the arc $(e^{i\theta};\ \theta_1 \leqq \theta \leqq \theta_2)$. Let $\Delta$ be any connected component of the inverse image of $(c)\colon |w - \alpha| < \varrho$ in the domain $D_0$; the existence of $\Delta$ is evident since $\alpha$ is a cluster value of $f(z)$ at $z_0$. Clearly $\Delta$ is not compact in $D_0$. We map the universal covering surface $\tilde{\Delta}$ of $\Delta$ conformally upon the unit disc $|\zeta| < 1$ in a one-to-one manner. Denote by $z = z(\zeta)$ the mapping function. Considering the two bounded analytic functions $z = z(\zeta)$ and $w = w(\zeta) = f(z(\zeta))$ in $|\zeta| < 1$, we denote by $E_\zeta$ the set of $\zeta = e^{i\varphi}$ such that the both radial limits $z(e^{i\varphi})$ and $w(e^{i\varphi})$ exist and such that $|w(e^{i\varphi}) - \alpha| < \varrho$. Consider first the case where there exists a point $e^{i\varphi} \in E_\zeta$ such that $z(e^{i\varphi})$ coincides with some $z = e^{i\theta} \in \mathscr{C}E$. Then there exists an asymptotic path $\Lambda$ terminating at $z = e^{i\theta}$ along which $f(z)$ converges to the asymptotic value $\beta = w(e^{i\varphi})$. If $\beta = \alpha$, our assertion is true. If $\beta \neq \alpha$, applying a classical theorem on cluster sets[2], we see that there is an asymptotic path in $D$ ending at $e^{i\theta}$ along which $f(z) \to \alpha$. In the other case, we have seen, in the proof of Theorem 9, that $(w(\zeta) - \alpha)/\varrho$ is a function of class $(U)$. Since $\alpha$ is an exceptional value of $w(\zeta)$, $\alpha$ must be a radial limit of $w(\zeta)$. Accordingly, $\alpha$ is an asymptotic value of $f(z)$ at a point $z = e^{i\theta}$ belonging to $E$.

**Theorem 11.** *If $\Omega = C_D(f, z_0) - C^*_{\Gamma-E}(f, z_0)$ is not empty, then the range of values $R_D(f, z_0)$ of $f(z)$ at $z_0$ covers $\Omega$, with a possible exceptional set of capacity zero* (NOSHIRO [11]).

---

[1] It is easy to see that Theorem 7 remains valid in the case of a function of generalized class $(U)$.

[2] Cf. Theorem 1, § 3, II and Theorem 4, § 4, II.

*Proof.* Suppose that $\mathscr{C}R_D(f, z_0) \cap \Omega$ is of positive capacity. Then there exists a positive number $r_0$ such that the complement of the set of values taken by $f(z)$ in the intersection of $D$ and $|z - z_0| < r_0$ contains a Borel set $e$ of positive capacity in $\Omega$. We can find a point $\alpha$ of $e$ such that for all positive numbers $\eta$, $U(\alpha, \eta) \cap e$ is of positive capacity. Select a positive number $r$ ($< r_0$) such that $\alpha$ is outside $M_r$ and denote by $\varrho$ a positive number less than the distance of $\alpha$ from $M_r$. By Theorem 10, it is possible to find a point $z_0' = e^{i\theta_0'}$ (distinct from $z_0$ or not) arbitrarily near $z_0$ such that $\alpha$ is an asymptotic value of $f(z)$ at $z_0'$. In the case that $z_0' \neq z_0$, we see that $z_0'$ belongs to $E$.[1] Accordingly, we can define $C^*_{\Gamma-E}(f, z_0')$ in the same manner as $C^*_{\Gamma-E}(f, z_0)$. Then $C_D(f, z_0') - C^*_{\Gamma-E}(f, z_0') \supset U(\alpha, \varrho)$. The reasoning in the proof of Theorem 10 shows that $R_D(f, z_0')$ covers the disc $U(\alpha, \varrho)$ with a possible exceptional set of capacity zero. Thus we arrive at a contradiction.

**Theorem 12.** *Let $H$ be a given closed set in the $w$-plane, $\Omega_1$ a connected component of the complement $\mathscr{C}H$ of $H$ with respect to the $w$-plane, and $\alpha$ an accessible boundary point of $\Omega_1$. Let the cluster set $C_{\Lambda_\theta}(f, e^{i\theta})$ be included in $H$ for every $e^{i\theta} \in \mathscr{C}E$, and let $C_D(f, z_0) \cap \Omega_1 \neq \emptyset$. Furthermore, let the set $\Xi$ of $e^{i\theta} \in \mathscr{C}E$ with $\alpha \in C_{\Lambda_\theta}(f, e^{i\theta})$ be of linear measure zero. Then, if $\alpha$ is an exceptional value of $f(z)$ in a neighborhood of $z_0$, $\alpha$ is an asymptotic value of $f(z)$ arbitrarily near $z_0$* (Noshiro [11]).

*Proof.* It is obvious that $C^*_{\Gamma-E}(f, z_0) \subset H$. Since $C_D(f, z_0) - C^*_{\Gamma-E}(f, z_0)$ is an open set, $\Omega_1 \subset C_D(f, z_0)$. We select a sufficiently small positive number $r$ such that $f(z) \neq \alpha$ in the intersection of $D$ and $(K)$: $|z - z_0| < r$. Next we select two points $z_1 = e^{i\theta_1}$, $z_2 = e^{i\theta_2}$ ($\theta_1 < \theta_0 < \theta_2$) belonging to $(\mathscr{C}E - \Xi) \cap (K)$. Using the same method as before, we construct a cross-cut $L$, consisting of two last parts of $\Lambda_{\theta_1}$, $\Lambda_{\theta_2}$ and a segment, of $D$ inside $(K)$. We may assume that there exists a positive number $\varrho_1$ such that $|f(z) - \alpha| \geqq \varrho_1 > 0$ on $L$. Let $\varrho$ be a positive number less than $\varrho_1$. We draw a circle $c$: $|w - \alpha| = \varrho$ and denote by $(c)$ its interior. As $\alpha$ is an accessible boundary point, we can draw a simple closed curve $Q$ inside $(c)$ which passes through $\alpha$ and which lies in $\Omega_1$ except for the point $\alpha$ and whose interior $G$ is contained in $\Omega_1$. Consider a connected component $\Delta$ of the inverse image $f^{-1}(G)$ of $G$ in the domain $D_1$ bounded by $L$ and a circular arc ($e^{i\theta}$; $\theta_1 \leqq \theta \leqq \theta_2$). In the present case, $\Delta$ is simply connected since $\alpha$ is an exceptional value. We map $\Delta$ conformally upon the unit disc $|\zeta| < 1$. Let $z = z(\zeta)$ be the mapping function. Consider two bounded regular functions $z = z(\zeta)$ and $w = w(\zeta) = f(z(\zeta))$, and denote by $E_\zeta$ the

[1] Denote by $\Lambda$ the path ending at $z_0'$ for which $f(z)$ has the asymptotic value $\alpha$. Assume that $z_0' \in \mathscr{C}E$. Then, applying Theorem 8, § 4, II to a Jordan domain $D'$ bounded by two suitable last parts of $\Lambda$, $\Lambda_{\theta_0'}$ and a suitable simple arc in $D$ and $z_0'$, we see that $R_{D'}(f, z_0')$ covers $\Omega_\alpha - (\alpha)$ with two possible exceptions, $\Omega_\alpha$ denoting the component of $\mathscr{C}C_{\Lambda_{\theta_0'}}(f, z_0')$, with respect to the $w$-plane, which contains the point $\alpha$.

set of $\zeta = e^{i\varphi}$ such that both radial limits $z(e^{i\varphi})$ and $w(e^{i\varphi})$ exist and such that $w(e^{i\varphi})$ lies in $G$. First consider the case there exists a point $e^{i\varphi} \in E_\zeta$ such that $z(e^{i\varphi})$ coincides with some $z = e^{i\theta} \in \mathscr{C}E$. Then we have an asymptotic path ending at $z = e^{i\theta}$ along which $f(z)$ converges to the asymptotic value $\beta = w(e^{i\varphi})$ lying in $G$. There are two cases depending on whether $C_{\Lambda_\theta}(f, e^{i\theta})$ contains $\alpha$ or not. In case $\alpha \notin C_{\Lambda_\theta}(f, e^{i\theta})$, by a classical theorem on cluster sets, we see that there exists an asymptotic path in $D$ ending at $z = e^{i\theta}$ with its asymptotic value $\alpha$. Thus our assertion holds. In case $\alpha \in C_{\Lambda_\theta}(f, e^{i\theta})$, $e^{i\theta}$ must belong to the set $\Xi$ of measure zero, by definition of $\Xi$. Therefore, we have only to consider the remaining case that $z(e^{i\varphi})$ belongs to $E \cup \Xi$ for every $e^{i\varphi} \in E_\zeta$. As $E \cup \Xi$ is of linear measure zero, $E_\zeta$ must be of linear measure zero. Now, we map $G$ by $W = \Phi(w)$ conformally upon the unit disc $|W| < 1$ in a one-to-one manner. Then, the composed function $W = W(\zeta) = \Phi[f(z(\zeta))]$ becomes a function of class $(U)$. Hence, $\Phi(\alpha)$ is a radial limit $W(e^{i\varphi})$ of $W(\zeta)$. The image of the radius with end point $\zeta = e^{i\varphi}$ by $z = z(\zeta)$ terminates at some point $z = e^{i\theta}$ on $\Gamma$, as $\Xi$ is of linear measure zero. Thus $\alpha$ is an asymptotic value of $f(z)$ at the point $z = e^{i\theta}$.

Remark. Theorem 9 contains a theorem of Ohtsuka [1]. This theorem is also closely related to Collingwood-Cartwright's main theorem in the small (cf. § 2, III). Theorem 10 contains a theorem of Lohwater ([2], Theorem 6, p. 250). Theorem 11 is an analogue of Theorem 5, § 4, II and contains another theorem of Lohwater ([2], Theorem 8, p. 251). In Theorem 12, if we take as $\Lambda_\theta$ the radius with end point $z = e^{i\theta}$, and if $f(z)$ is a meromorphic function of bounded type, the set $\Xi$ is necessarily of linear measure zero, by Nevanlinna's extension of the Fatou-Riesz theorem. Accordingly, Theorem 12 is considered as an extension of theorems of Lohwater [3], Lehto [3] and Storvick [1][1].

**4. Applications.** Let $w = f(z)$ be meromorphic in the unit circle $|z| < 1$. Using the same notations in Paragraph 3, we start with the following

**Theorem 13.** *Suppose that for every* $e^{i\theta} \in \mathscr{C}E = A - E$, $C_{\Lambda_\theta}(f, e^{i\theta})$ *lies on the circumference* $\Gamma_w$: $|w| = 1$ *and that* $C_D(f, z_0) = C^*_{\Gamma - E}(f, z_0)$. *Then,* $w = f(z)$ *is regular at* $z = z_0$.

*Proof.* Obviously, $f(z)$ is regular and bounded in the intersection of $|z| < 1$ with a neighborhood of $z_0$; there is an open arc $A_1$ of $A$, containing $z_0$ in its interior, such that there exists the radial limit $f(e^{i\theta})$ for almost every $e^{i\theta} \in A_1 - E$. Since there are at most countably many ambiguous points of $f(z)$ on $|z| = 1$, we have $|f(e^{i\theta})| = 1$ for almost every point $e^{i\theta}$ of $A_1 - E$. Hence, $f(z)$ is a function of generalized class $(U)$. Thus $f(z)$ is regular at $z_0$.

[1] As for related results, cf. Ohtsuka [9, 15], Lohwater [3—8], Storvick [2], Lehto [7, 9].

As an immediate consequence of Theorems 9, 10, 13, we have

**Theorem 14.** *Suppose that for every* $e^{i\theta} \in \mathscr{C}E = A - E$, $C_{\Lambda_\theta}(f, e^{i\theta})$ *lies on* $\Gamma_w$: $|w| = 1$ *and that* $w = f(z)$ *has a singularity at* $z = z_0$. *Then,* $\Omega = C_D(f, z_0) - C^*_{\Gamma - E}(f, z_0)$ *contains at least one of* $|w| < 1$ *and* $|w| > 1$. *If, further,* $\alpha \notin R_D(f, z_0)$ *and* $\alpha \in \Omega$, *then* $\alpha$ *is an asymptotic value of* $w = f(z)$ *arbitrarily near* $z_0$.

Example (Lohwater [2]). The function

$$u(r, \theta) = \Re\left[i \frac{2z}{(1-z)^2}\right] = \frac{-2r(1-r^2)\sin\theta}{(1+r^2-2r\cos\theta)^2}$$

is harmonic in $|z| < 1$ and for all $\theta$, $\lim_{r\to 1} u(r, \theta) = 0$. If $v(r, \theta)$ is its conjugate, the function[1]

$$g(z) = e^{u+iv} = \exp\left[i \frac{2z}{(1-z)^2}\right]$$

is regular, $\neq 0$ in $|z| < 1$, and, along all radii, $|g(re^{i\theta})| = e^{u(r,\theta)} \to 1$ as $r \to 1$. Therefore, 0 and $\infty$ are not radial limits. On the other hand, along the upper half of the oricycle $r = \cos\theta$ $(0 < \theta < \pi/2)$, $u(r, \theta) \to -\infty$, as $r \to 1$, whence $g(z) \to 0$. Similarly, on the lower half of the oricycle $r = \cos\theta$ $(-\pi/2 < \theta < 0)$, $u(r, \theta) \to +\infty$ as $r \to 1$, whence $g(z) \to \infty$. Obviously $g(z)$ is not of bounded type, since $\int_{-\pi}^{\pi} |u(r, \theta)|\, d\theta$ is not bounded in $0 < r < 1$.

If we adopt as $\Lambda_\theta$ the radius with end point $e^{i\theta}$, then the following theorem is an immediate consequence of Theorem 12.

**Theorem 15.** *Let* $f(z)$ *be a meromorphic function of bounded type in* $D$: $|z| < 1$ *and let* $\lim_{r\to 1} |f(re^{i\theta})| = 1$ *almost everywhere on an open arc* $A$ *of* $|z| = 1$. *If* $f(z)$ *has a singularity at a point* $z_0$ *of* $A$, *then every value of modulus* 1 *which does not belong to* $R_D(f, z_0)$ *is an asymptotic value of* $f(z)$ *at some point of each arc of* $A$ *containing the point* $z_0$ (Lohwater [3, 4]).

Lohwater ([4], p. 156) raised a question whether this theorem holds in the case of a meromorphic function of unbounded type. We now discuss on this problem, although it is difficult to give a complete answer. With a slight modification of the proof of Theorem 12, we can prove

**Theorem 16.** *Let* $f(z)$ *be meromorphic in* $|z| < 1$. *Suppose that* $C_{\Lambda_\theta}(f, e^{i\theta})$ *lies on* $\Gamma_w$: $|w| = 1$ *for almost every* $e^{i\theta}$ *belonging to an open arc* $A$ *of* $\Gamma$ *containing* $z_0 = e^{i\theta_0}$, *where* $\Lambda_\theta$ *is an arbitrary curve in* $|z| < 1$ *ending at* $e^{i\theta}$. *Let* $f(z)$ *have a singularity at* $z_0$. *If* $f(z)$ *has an exceptional value* $\alpha$ $(|\alpha| = 1)$ *in a neighborhood of* $z_0$, *then* $\alpha$ *is an asymptotic value along a path terminating at some point arbitrarily near* $z_0$, *under an additional condition that*

(i) $\mathscr{E} = \{e^{i\theta};\, \alpha \notin C_{\Lambda_\theta}(f, e^{i\theta})\}$ *is everywhere dense on* $A$.

[1] Cf. Lehto [8] on the construction of such functions.

Remark. Suppose that $f(z)$ has three exceptional values in a neighborhood of $z_0$. Then it is known that the radial limit $f(e^{i\theta})$ exists at a dense subset of $A$[1]. Hence, in Theorem 16, the additional condition (i) can be replaced by the condition that

(ii) $f(z)$ has three exceptional values in a neighborhood of $z_0$.

Furthermore, if we make use of Theorem 10, then in the same theorem the condition (i) can be also replaced by the condition

(iii) there exists a point $\beta$ such that $\beta \notin R_D(f, z_0)$, $|\beta| \neq 1$.

From the preceding remark follows

**Theorem 17.** *Let $f(z)$ be regular in $D$: $|z| < 1$. Suppose that $C_{A_\theta}(f, e^{i\theta})$ lies on $\Gamma_w$: $|w| = 1$ for almost every $e^{i\theta}$ belonging to $A$. Let $f(z)$ have a singularity at $z_0 = e^{i\theta_0}$. If $f(z)$ has an exceptional value $\alpha$ ($|\alpha| = 1$) in a neighborhood of $z_0$, then $\alpha$ is an asymptotic value of $f(z)$ at some point of each arc of $A$ containing $z_0$.*

Now we treat the case of radial cluster sets, by applying an idea of LOHWATER [5]. Let $w = f(z)$ be meromorphic in $D$: $|z| < 1$. Suppose that $\lim_{r\to 1} |f(re^{i\theta})| = 1$ almost everywhere on an arc $A$ and that $\Sigma(1 - |a_k|) < \infty$, where $a_k$ $(k = 1, 2, \ldots)$ are zeros of $f(z)$ in $|z| < 1$. Assume that $C_D(f, e^{i\theta})$ is *total*[2] for every $z = e^{i\theta} \in A$. Consider the function $\varphi_\varrho(z) \equiv \equiv f(z)/B_\varrho(z)$, in the intersection of $D$ with a neighborhood $V_\varrho$: $|z - e^{i\theta}| < < \varrho$, where $B_\varrho(z)$ is a Blaschke product formed by zeros of $f(z)$ in $D \cap V_\varrho$. By hypothesis, $\varphi_\varrho(z)$ has a singularity at $e^{i\theta}$; otherwise, $f(z)$ would be bounded in a neighborhood of $e^{i\theta}$. Evidently, $\lim_{r\to 1} |\varphi_\varrho(re^{i\theta'})| = 1$ for almost every $e^{i\theta'}$ of an open arc of $A$ containing the point $e^{i\theta}$. There are two cases depending on whether $C_D(\varphi_\varrho, e^{i\theta})$ contains $w = 0$ or not. Consider first the case in which $C_D(\varphi_\varrho, e^{i\theta})$ does not contain $w = 0$. Then $|\varphi_\varrho(z)| > \delta > 0$ in $D \cap V_{\varrho'}$ if $\varrho'(>0)$ is sufficiently small. Since $f(z) = B_\varrho(z)/\frac{1}{\varphi_\varrho(z)}$, $f(z)$ can be written in the form of the quotient of bounded regular functions in $D \cap V_{\varrho'}$; in other words, $f(z)$ is locally of bounded type. Hence we can apply the Nevanlinna-Fatou-Riesz theorem. Next, consider the case in which $C_D(\varphi_\varrho(z), e^{i\theta})$ contains $w = 0$. Since $w = 0$ is an exceptional value for $\varphi_\varrho(z)$, there exists an asymptotic path $\mathscr{L}$, with asymptotic value 0, terminating at a point of $A$ arbitrarily near $e^{i\theta}$. From $f(z) = B_\varrho(z) \cdot \varphi_\varrho(z)$, we see that $w = f(z)$ converges to $w = 0$ along $\mathscr{L}$.

Using these preliminary discussions and Theorem 16, we have

**Theorem 18.** *Let $f(z)$ be meromorphic in $|z| < 1$. Suppose that* $\lim_{r\to 1} |f(re^{i\theta})| = 1$ *almost everywhere on an arc $A$ $(e^{i\theta}: \theta_1 < \theta < \theta_2)$ of $|z| = 1$. Let $f(z)$*

[1] COLLINGWOOD-CARTWRIGHT [1].

[2] If $C_D(f, e^{i\theta})$ coincides with the whole $w$-plane, it is called total.

*have a singularity at* $z_0 = e^{i\theta_0}$ $(\theta_1 < \theta_0 < \theta_2)$. *If* $f(z)$ *has an exceptional value* $\alpha\,(|\alpha| = 1)$ *at* $z_0$, *then* $\alpha$ *is an asymptotic value arbitrarily near* $z_0$, *provided that* $\Sigma(1 - |b_k|) < \infty$ *where* $b_k$ $(k = 1, 2, \ldots)$ *are* $\beta$*-points of* $f(z)$ *and* $|\beta| \neq 1$[1].

## § 2. Boundary theorems of Collingwood and Cartwright

The object of this section is to state main theorems of Collingwood and Cartwright [1] for functions meromorphic in the unit circle and related theorems.

1. Let $w = f(z)$ be meromorphic in the unit circle $D$: $|z| < 1$. We define the cluster set $C(f)$, the range of values $R(f)$ and the asymptotic set $A(f)$ (in the large) as follows:

(i) The **Cluster Set** $C(f)$. $\alpha \in C(f)$ if there exists a sequence $\{z_n\}$, $|z_n| < 1$, such that $\lim\limits_{n\to\infty} |z_n| = 1$ and $\lim\limits_{n\to\infty} f(z_n) = \alpha$. Evidently $C(f)$ is a single point or a continuum.

(ii) The **Range of Values** $R(f)$. $\alpha \in R(f)$ if there exists a sequence $\{z_n\}$, $|z_n| < 1$, such that $\lim\limits_{n\to\infty} |z_n| = 1$ and $f(z_n) = \alpha$ for all $n$. Obviously, $R(f)$ is a $G_\delta$ set.

(iii) The **Asymptotic Set** $A(f)$. $\alpha \in A(f)$ if there is a continuous curve $\Lambda : z = z(t)$, $0 \leqq t < 1$, such that $|z(t)| < 1$; $\lim\limits_{t\to 1} |z(t)| = 1$ and $\lim\limits_{t\to 1} f(z(t)) = \alpha$. We call $\alpha$ an asymptotic value of $f(z)$ with asymptotic path $\Lambda$. Here we do not assume that $\Lambda$ terminates at a single point on $\Gamma$: $|z| = 1$. More precisely, if we call the limiting set on $\Gamma$ of the asymptotic path $\Lambda$ the "end" of $\Lambda$, then the end of $\Lambda$ is either a single point on $\Gamma$ or a closed arc of $\Gamma$ which may be the whole circumference $\Gamma$. We say that if $\alpha \in A(f)$, $\alpha$ belongs to $A_p(f)$ or $A_a(f)$, according as the end of $\Lambda$ is a single point or not. Obviously, $A(f) = A_p(f) \cup A_a(f)$.

We now state shortly some results concerning the characterization of $C(f)$, $R(f)$ and $A(f)$. Every cluster set $C(f)$ is a continuum which may be degenerate to one point but not every continuum is a cluster set[2]. A counter-example of Rudin [2] is as follows: Let $K$ be the continuum consisting of a spiral $S: r = \dfrac{\theta}{\pi + \theta}$, $\pi \leqq \theta < \infty$, the unit circumference $\Gamma$ and a segment $I$: $1 \leqq x \leqq 2$, $y = 0$. Let us call a set $E$, in the complex plane, a $\Delta$*-set*, if $E = \bigcap\limits_{n=1}^{\infty} G_n$, where $G_n \supset G_{n+1}$ and each $G_n$ is a connected

[1] We may assume that $\beta$ is equal to 0; otherwise, make a transformation $W = \dfrac{w - \beta}{1 - \bar{\beta} w}$.

[2] Compare it with the following theorem of Gross: Given any continuum $K$ and any $\theta$, there is a function $f(z)$ meromorphic in $|z| < 1$, such that $K = C(f, e^{i\theta})$ (Gross [1], p. 20).

open set. Obviously $R(f)$, as well as $R_D(f, e^{i\theta})$, is a $\Delta$-set. RUDIN [2] has proved: *If $E$ is a $\Delta$-set, then there exists a meromorphic function in $|z| < 1$, such that $R(f) = E$; more precisely, $R_D(f, 1) = E$ and $R_D(f, e^{i\theta}) = \emptyset$ if $\theta \neq 0$. Furthermore, every closed set is a $\Delta$-set but there is an open set which is not a $\Delta$-set.* EGGLESTON [1] has succeeded in characterizing the range of values $R(f)$ by means of a process, which involves a transfinite number of operations. On the other hand, *the characterization of the cluster set $C(f)$ seems still open.* As for the asymptotic set $A(f)$, it is well-known that *given any analytic set $E$, there exists a meromorphic function in $|z| < 1$ such that $A(f) = E$* (KIERST [1])[1].

Remark. In this section, we use the following notations. Let $E$ be an arbitrary set in the complex plane. We denote by $\overline{E}$, $\mathscr{C}E$, $\mathscr{F}E$ and $\mathscr{I}E$ the closure of $E$, the complement of $E$ with respect to the whole complex plane, the boundary of $E$ and the interior of $E$ respectively.

**2. Collingwood-Cartwright's main theorem in the large.**

**Lemma 1.** *Let $w = f(z)$ be a non-constant meromorphic function in $D$: $|z| < 1$ which is bounded in the intersection of $D$ with a neighborhood $U(z_0, r)$: $|z - z_0| < r$ of a point $z_0$ on $\Gamma$: $|z| = 1$. Then, $A_{\mathfrak{p}}(f)$ is of positive linear measure.*

*Proof.* By Fatou's theorem, $f(z)$ has an angular limit at almost every point of an arc $(e^{i\theta}; \theta_1 \leqq \theta \leqq \theta_2)$, where $\theta_1 < \theta_2 < \theta_2$ and $z_0 = e^{i\theta_0}$, in $U(z_0, r)$. We may assume, without loss of generality, that $f(z)$ has distinct angular limits $w_1 = f(e^{i\theta_1})$, $w_2 = f(e^{i\theta_2})$ by Riesz' theorem. Choose a circular arc $K$ in $D \cap U(z_0, r)$ with end points $e^{i\theta_1}$, $e^{i\theta_2}$ such that $f'(z) \neq 0$ on $K$. Then, the image $\mathscr{L}$ of $K$ by $w = f(z)$ is a regular analytic curve in the $w$-plane with end points $w_1$, $w_2$. Consider any straight line $g$ which is perpendicular to the segment $S$ connecting $w_1$ and $w_2$, and intersects $S$ between $w_1$ and $w_2$. Obviously, since any analytic function possesses at most a countable number of algebraic singularities, we may suppose that the inverse of $w = f(z)$ has no algebraic singularity on $g$. Now, consider the inverse image $f^{-1}(g)$ of $g$ in the domain $D_1$ bounded by $K$ and the arc $\mathfrak{A}(e^{i\theta}; \theta_1 \leqq \theta \leqq \theta_2)$. Then, there exists at least one cross-cut $\Lambda$ of $D_1$ starting at a point of $K$ and terminating at a point $\zeta$ of $\mathfrak{A}$, since $\mathscr{L}$ intersects $g$ an odd number of times. Evidently, $w = f(z)$ has an asymptotic value, lying on $g$, with the asymptotic path $\Lambda$. Consequently, $A_{\mathfrak{p}}(f)$ is of positive measure.

**Lemma 2.** *Let $f(z)$ be meromorphic in $D$: $|z| < 1$. If $\alpha \in \mathscr{C}\overline{A}(f) \cap C(f)$, then $\alpha$ is an interior point of $R(f)$.*

*Proof.* By hypothesis, there exists a positive number $\varrho_0$ such that $U(\alpha, \varrho_0) \cap A(f) = \emptyset$, where $U(\alpha, \varrho_0)$ denotes the $\varrho_0$-neighborhood of $\alpha$.

[1] For a related theorem, see Theorem 14, § 4, III (BAGEMIHL-SEIDEL [3]). Cf. also HEINS [4] in the case of an entire function.

For any positive number $\varrho < \varrho_0$, consider the inverse image $f^{-1}(G)$ in $D$ of $G$: $|w - \alpha| < \varrho$. If the frontier of $f^{-1}(G)$ contains infinitely many closed analytic contours, then evidently the covering surface $\Phi$ generated by $w = f(z)$ covers every point of $|w - \alpha| = \varrho$ infinitely often. On the other hand, if the frontier of $f^{-1}(G)$ contains only a finite number of closed analytic contours, then there exists at least one non-compact connected component $\Delta$ of $f^{-1}(G)$. We note that the frontier of $\Delta$ must contain at least one open analytic contour $\Lambda$; otherwise, $\Delta$ would contain an annulus $1 - \varepsilon < |z| < 1$ and by Fatou's theorem we would have $A_{\mathfrak{p}}(f) \cap \overline{G} \neq \theta$ which contradicts the relation $A(f) \cap U(\alpha, \varrho_0) = \theta$. Let $z_\varrho$ be any point on $\Lambda$ and $e(w, w_\varrho)$ be the corresponding functional element of the inverse of $w = f(z)$. Then, the continuation of algebraic character of $e(w, w_\varrho)$ along $|w - \alpha| = \varrho$ is possible indefinitely. Accordingly, the covering surface $\Phi$ covers every point of $|w - \alpha| = \varrho$ infinitely often. Thus we see that $U(\alpha, \varrho_0) \subset R(f)$ except perhaps the point $\alpha$. We now show that $\alpha$ also belongs to $R(f)$. Let $\alpha'$ be any point in $0 < |w - \alpha| < \varrho_0/2$ and put $\varrho' = |\alpha' - \alpha|$. Then, obviously $A(f) \cap U(\alpha', \varrho_0/2) = \theta$ and $\alpha' \in C(f)$. Consequently, by the foregoing argument for $|w - \alpha'| = \varrho'$, $\alpha$ must belong to $R(f)$. Hence we have proved $U(\alpha, \varrho_0) \subset R(f)$.

**Lemma 3.** *If $f(z)$ is meromorphic in $|z| < 1$, then*

$$\mathscr{F}R(f) \cup \mathscr{F}C(f) = \overline{\mathscr{C}R}(f) \cap C(f).$$

*Proof.* Evidently

$$\mathscr{F}R(f) \subset \overline{\mathscr{C}R}(f) \quad \text{and} \quad \mathscr{F}R(f) \subset C(f), \tag{1}$$

since $R(f) \subset C(f)$. On the other hand,

$$\mathscr{F}C(f) \subset \overline{\mathscr{C}C}(f) \subset \overline{\mathscr{C}R}(f), \tag{2}$$

since $\mathscr{I}R(f) \subset \mathscr{I}C(f)$.

From (1) and (2),

$$\mathscr{F}R(f) \cup \mathscr{F}C(f) \subset \overline{\mathscr{C}R}(f) \cap C(f). \tag{3}$$

Next, we prove the opposite inclusion of (3). Assume that $\alpha \in \overline{\mathscr{C}R}(f) \cap$ $\cap C(f)$ and $\alpha \notin \mathscr{F}R(f) \cup \mathscr{F}C(f)$. Then $\alpha \in \mathscr{I}C(f)$. It is easily proved that $R(f)$ is dense in $\mathscr{I}C(f)$[1]; hence $\alpha \in \overline{R}(f)$. However $\alpha \notin \mathscr{F}R(f)$. Consequently $\alpha$ must belong to $\mathscr{I}R(f)$ which is contradictory to the assumption.

Now we prove Collingwood-Cartwright's main theorem in the large.

**Theorem 1.** *If $f(z)$ is non-constant and meromorphic in $|z| < 1$, then the following relations are satisfied:*

(i) *If $A(f)$ is unrestricted,*

$$\mathscr{F}R(f) \cup \mathscr{F}C(f) = \overline{\mathscr{C}R}(f) \cap C(f) \subset \overline{A}(f); \tag{4}$$

[1] Cf., for example, COLLINGWOOD and CARTWRIGHT [1], p. 109 and NOSHIRO [2], p. 230.

(ii) *If $A(f)$ is of linear measure zero,*

$$\mathscr{C}R(f) \subset A(f) \tag{5}$$

(COLLINGWOOD-CARTWRIGHT [1]).

*Proof.* To prove (i) we use Lemma 2. By that lemma

$$\mathscr{I}R(f) \supset \mathscr{C}\bar{A}(f) \cap C(f)$$

and therefore, taking complements,

$$\overline{\mathscr{C}R}(f) \subset \bar{A}(f) \cup \mathscr{C}C(f);$$

hence

$$\overline{\mathscr{C}R}(f) \cap C(f) \subset \bar{A}(f).$$

The complete relation (4) follows from Lemma 3.

To prove (ii), we use Lemma 1. By that lemma, $C(f)$ is the whole $w$-plane. Applying the method used in the proof of Lemma 2, we see that every circumference $c$: $|w - \alpha| = \varrho$, except perhaps for a set of positive numbers $\varrho$ of measure zero, is contained in $R(f)$. Next note that almost every radius $\Lambda_\theta$: $w = \alpha + te^{i\theta}$, $0 < t \leqq \varrho$ of $c$ does not meet $A(f)$ and that every analytic function has at most enumerably many algebraic singularities. Consequently, we can find a point $\beta = \alpha + \varrho\, e^{i\theta}$ with the following properties:

(6) The inverse of $w = f(z)$ has infinitely many regular functional elements $e_n(w, \beta)$ $(n = 1, 2, \ldots)$ with center $\beta$.

(7) The continuation (of rational character) of every element $e_n(w, \beta)$ along a fixed radius $\Lambda_\theta$: $w = \alpha + t\, e^{i\theta}$, $0 < t \leqq \varrho$ is possible.

Suppose that $\alpha \notin R(f)$. Then it is obvious that there exists at least one element $e_n(w, \beta)$ whose continuation along the radius $\Lambda_\theta$ defines a transcendental singularity at $w = \alpha$. Thus it is proved that $\alpha \in A(f)$[1].

Remark. In the general case, the relation (4) is best possible in the sense that $\bar{A}(f)$ cannot be replaced by $A(f)$. Let $\Delta_w$ be a multiply-connected domain in the $w$-plane whose boundary $\Gamma_w$ has an inaccessible boundary point $\omega$ of $\Delta_w$. Let $w = g(z)$ be a function which maps the universal covering surface $\tilde{\Delta}_w$ of $\Delta_w$ onto $|z| < 1$ in a one-to-one conformal manner. Then, by the theory of conformal mappings, $R(g) = \Delta_w$, $C(g) = \bar{\Delta}_w$; and $A(g)$ consists of all accessible boundary points. Consequently, $\omega \in \mathscr{C}R(g) \cap C(g) \cap \mathscr{C}A(g)$. For example, denote by $s_n$ the segment $u = 1/n$, $-1 \leqq v \leqq +1$, where $w = u + iv$ and by $\Delta_w$ the domain obtained by cutting the $w$-plane along all the segments for $n = \pm 1, \pm 2, \ldots$ so that the boundary $\Gamma_w$ of $\Delta_w$ consists of all the $s_n$ and

[1] If $A(f)$ is of linear measure zero, then, for every point $z_0$ of $\Gamma$: $|z| = 1$, the cluster set $C_D(f, z_0)$ is the whole $w$-plane by Lemma 1. Hence the unit circumference $\Gamma$ is the natural boundary for $w = f(z)$. Obviously the inverse $z = \varphi(w)$ of $w = f(z)$ has Gross' property and so Iversen's property (cf. § 2, I). The fact $\mathscr{C}R(f) \subset A(f)$ is also an immediate consequence of these properties.

the segment $s_\infty$ defined by $u = 0, -1 \leqq v \leqq +1$. Obviously all internal points of $s_\infty$ are inaccessible boundary points of $\Delta_w$. This example is due to COLLINGWOOD and CARTWRIGHT [1]. Cf. also BAGEMIHL and SEIDEL [7], OHTSUKA [9].

**3. Collingwood-Cartwright's theorem in the small**[1]. Let $w = f(z)$ be meromorphic in the unit circle $D$: $|z| < 1$. We denote by $\Gamma$ the unit circumference $|z| = 1$. As stated in § 1, I, we associate with every point $z_0 = e^{i\theta_0}$ of $\Gamma$ the cluster set $C_D(f, z_0)$, the boundary cluster set $C_\Gamma(f, z_0)$, the range of values $R_D(f, z_0)$ and the asymptotic set $A_D(f, z_0)$. In this paragraph, for the sake of simplicity, we write $C(f, z_0)$, $R(f, z_0)$ and $A(f, z_0)$ instead of $C_D(f, z_0)$, $R_D(f, z_0)$ and $A_D(f, z_0)$. Obviously $C(f, z_0)$ is a single point or a continuum.

To state Collingwood-Cartwright's main theorem in the small, we need to adopt the following additional notations and definitions. We say that

$$\alpha \in A(f, \theta_1 < \theta < \theta_2) \quad \text{or} \quad \alpha \in A(f, \theta_1 \leqq \theta \leqq \theta_2),$$

if there is an asymptotic path on which $f(z)$ converges to $\alpha$ and whose end[2] is contained in the open arc $z = e^{i\theta}$, $\theta_1 < \theta < \theta_2$ or the closed arc $z = e^{i\theta}$, $\theta_1 \leqq \theta \leqq \theta_2$, respectively. The intersection of the sets $A(f, |\theta - \theta_0| < \eta)$, which we denote by

$$\chi(f, e^{i\theta_0}) = \bigcap_\eta A(f, |\theta - \theta_0| < \eta), \tag{8}$$

plays the same rôle in the theory in the small as $A(f)$ in the theory in the large. We write

$$\chi^*(f, e^{i\theta_0}) = \bigcap_\eta \bar{A}(f, |\theta - \theta_0| < \eta).^3 \tag{9}$$

Obviously $\chi^*(f, e^{i\theta_0})$ is closely related to $C^*_{\Gamma - E}(f, z_0)$ introduced in § 1 and $\chi^*(f, e^{i\theta_0}) \subset C(f, e^{i\theta_0})$.

Let there be two sequences $\{z_n^{(1)}\}$ and $\{z_n^{(2)}\}$ such that $|z_n^{(1)}| < 1$, $\lim_{n\to\infty} z_n^{(1)} = e^{i\theta_1}$; $|z_n^{(2)}| < 1$, $\lim_{n\to\infty} z_n^{(2)} = e^{i\theta_2}$ where $\theta_1 \neq \theta_2$. If there is a sequence of continuous curves $\gamma_n$ joining $z_n^{(1)}$ to $z_n^{(2)}$ and contained in an annulus $1 - \varepsilon_n < |z| < 1$, where $\varepsilon_n > 0$, $\lim_{n\to\infty} \varepsilon_n = 0$, such that on $\gamma_n$ we have $|f(z) - \alpha| < \eta_n$ where $\lim_{n\to\infty} \eta_n = 0$, then we say that $\alpha$ belongs to $\Phi(f, e^{i\theta_0})$ for any $\theta_0$ satisfying $\theta_1 \leqq \theta_0 \leqq \theta_2$. It is well-known that if $f(z)$ is non-

[1] This theorem is closely related to results of SEIDEL [2], OHTSUKA [1, 9, 15], LOHWATER [2, 7, 8], LEHTO [3, 7], NOSHIRO [11].

[2] The end of an asymptotic path is either a single point or a closed arc of $\Gamma$.

[3] The set $\chi(f, e^{i\theta_0})$ may be empty even though $A(f, |\theta - \theta_0| < \eta)$ is not empty or any $\eta > 0$; but $\chi^*(f, e^{i\theta_0})$ is empty if and only if $A(f, |\theta - \theta_0| < \eta)$ is empty for ome $\eta > 0$.

constant and $\mathscr{C}R(f, e^{i\theta_0})$ contains at least three points, then $\Phi(f, e^{i\theta_0})$ is empty (Koebe-Gross' theorem[1]).

We prove now Collingwood-Cartwright's theorem in the small.

**Theorem 2.** *If $f(z)$ is non-constant and meromorphic in $D$: $|z| < 1$, then for any point $z_0 = e^{i\theta_0}$ of $\Gamma$: $|z| = 1$ the following relations are satisfied:*

(i) *If $\chi(f, z_0)$ is unrestricted,*

$$\mathscr{F}R(f, z_0) \cup \mathscr{F}C(f, z_0) \\ = \overline{\mathscr{C}R}(f, z_0) \cap C(f, z_0) \subset \chi^*(f, z_0) \cup \Phi(f, z_0)^2; \tag{10}$$

(ii) *If $A(f, |\theta - \theta_0| < \eta)$ is of linear measure zero for some $\eta > 0$,*

$$\mathscr{C}R(f, z_0) \subset \chi(f, z_0) \cup \Phi(f, z_0) \tag{11}$$

(COLLINGWOOD-CARTWRIGHT [1]).

*Proof.* (I) Consider first the case that $C(f, z_0)$ is *sub-total.*[3] Then, we can define the boundary cluster set $C^*_{\Gamma-E}(f, z_0)$, in § 1, by using the radial limits $f(e^{i\theta})$ of $f(z)$, such that

$$C^*_{\Gamma-E}(f, z_0) \subset \chi^*(f, z_0).$$

By Theorem 9, § 1, $C(f, z_0) - C^*_{\Gamma-E}(f, z_0)$ is an open set and therefore $C(f, z_0) - \chi^*(f, z_0)$ is open. By Theorem 10, § 1, we have

$$C(f, z_0) - \chi^*(f, z_0) \subset \mathscr{I}R(f, z_0); \quad \text{i. e.,} \\ C(f, z_0) \cap \mathscr{C}\chi^*(f, z_0) \subset \mathscr{I}R(f, z_0). \tag{12}$$

Accordingly

$$\overline{\mathscr{C}R}(f, z_0) \subset \chi^*(f, z_0) \cup \mathscr{C}C(f, z_0).$$

Thus we have

$$\overline{\mathscr{C}R}(f, z_0) \cap C(f, z_0) \subset \chi^*(f, z_0).$$

In this case, it is obvious that $\Phi(f, z_0)$ is empty.

Next consider the case that $C(f, z_0)$ is total. We divide the case into the following two cases, according as $\mathscr{C}R(f, z_0)$ contains at least three points or at most two points.

($\text{II}_1$) Assume that $\mathscr{C}R(f, z_0)$ contains at least three points. Then, $\Phi(f, z_0)$ is empty. We have only to show that $\mathscr{C}\chi^*(f, z_0) \subset R(f, z_0)$. Suppose, contrary to the assertion, that $\Omega = \mathscr{C}\chi^*(f, z_0) \cap \mathscr{C}R(f, z_0)$ is not empty. Let $\alpha$ be a point belonging to $\Omega$. We select a positive number $r$, sufficiently small, such that $f(z) \neq \alpha$ in $D$ for $|z - z_0| < r$ and $\alpha$ lies outside $\overline{A}(f, |\theta - \theta_0| < \eta)$, where $e^{i(\theta_0 \pm \eta)}$ denote the points of inter-

[1] See Paragraph 3, § 2, I. Cf. COLLINGWOOD-CARTWRIGHT [1], pp. 96—98.

[2] The proof of the relation

$$\mathscr{F}R(f, z_0) \cup \mathscr{F}C(f, z_0) = \overline{\mathscr{C}R}(f, z_0) \cap C(f, z_0)$$

is similar to that of Lemma 3.

[3] This means that $C(f, z_0)$ is not the whole $w$-plane.

section of $K$: $|z - z_0| = r$ with $\Gamma$[1]. Let $\varrho_1$ be the distance of $\alpha$ from $\bar{A}(f, |\theta - \theta_0| < \eta)$. Now, we can choose two points $\zeta_1 = e^{i\theta_1}$, $\zeta_2 = e^{i\theta_2}$ $(\theta_1 < \theta_0 < \theta_2)$ arbitrarily near $z_0$ such that the angular limits $f(e^{i\theta_1})$, $f(e^{i\theta_2})$ exist. Describe a circular arc $L$ through $\zeta_1$ and $\zeta_2$ in $D \cap (K)$, where $(K)$ denotes the disc $|z - z_0| < r$. Obviously, there exists a positive number $\varrho_2$ such that $|f(z) - \alpha| \geqq \varrho_2 > 0$ on $L$. Let $\varrho$ be a positive number less than $\min(\varrho_1, \varrho_2)$. Denote by $D_0$ the domain bounded by $L$ and the arc $(e^{i\theta}, \theta_1 \leqq \theta \leqq \theta_2)$. Since $\alpha$ is a cluster value of $f(z)$ at $z_0$, there exists a point $b$ in $D_0$ such that $\beta = f(b)$ lies in $(c)$: $|w - \alpha| < \varrho$.

Considering the counter-image, in $D_0$, of the segment $g$ joining $\beta$ to $\alpha$, we can find an asymptotic path starting at $b$ and terminating at some point of the arc $(e^{i\theta}, \theta_1 \leqq \theta \leqq \theta_2)$ along which $f(z)$ converges to a value lying on $g$. This is a contradiction.

($\mathrm{II}_2$) Assume that $\mathscr{C}R(f, z_0)$ contains at most two points. Let $\alpha \in \mathscr{C}R(f, z_0)$. Suppose that $\alpha \notin \Phi(f, z_0)$. We select a positive number $r$, sufficiently small, such that $f(z) \neq \alpha$ in $D$ for $|z - z_0| < r$. Let $\{z_n\}$ be a sequence of points in $D \cap (K)$, $(K)$ denoting the disc $|z - z_0| < r$, such that $z_n \to z_0$, $w_n = f(z_n) \to \alpha$ and $\arg(w_k - \alpha) \neq \arg(w_h - \alpha)$ for $k \neq h$. Consider the image $\Lambda_n$, in $D \cap (K)$, of the segment $g_n$ joining $w_n$ to $\alpha$ by the continuation of the corresponding element $e(w, w_n)$ along $g_n$. If there were infinitely many $\Lambda_n$ starting at $z_n$ and terminating at a point of $K$: $|z - z_0| = r$, then $\alpha$ would belong to $\Phi(f, z_0)$. Accordingly, there exists at least one $\Lambda_n$ whose end lies on $\Gamma$ along which $f(z)$ admits an asymptotic value $\beta$ lying on $g_n$. Since $r$ is taken to be arbitrarily small, $\alpha$ must belong to $\chi^*(f, z_0)$.

Now we treat the case (ii). Suppose that $A(f, |\theta - \theta_0| < \eta)$ is of linear measure zero for some fixed $\eta > 0$. By Lemma 1, $C(f, z_0)$ is total. To prove (ii), we slightly modify the foregoing arguments in ($\mathrm{II}_1$) and ($\mathrm{II}_2$) and make use of a method closely related to Gross' star theorem.

($\mathrm{III}_1$) Suppose that $\mathscr{C}R(f, z_0)$ contains at least three points. Let $\alpha \in \mathscr{C}R(f, z_0)$. We select a positive number $r$, sufficiently small, such that $f(z) \neq \alpha$ in $D$ for $|z - z_0| < r$ and $K$: $|z - z_0| = r$ intersects the arc $(e^{i\theta}, |\theta - \theta_0| < \eta)$ at two points. Suppose that there are two points $\zeta_1 = e^{i\theta_1}$, $\zeta_2 = e^{i\theta_2}$ $(\theta_1 < \theta_0 < \theta_2)$ arbitrarily near $z_0$ such that the angular limits $f(e^{i\theta_1})$, $f(e^{i\theta_2})$ exist and $f(e^{i\theta_1}) \neq \alpha$, $f(e^{i\theta_2}) \neq \alpha$. Let $L$ be a circular arc through $\zeta_1$ and $\zeta_2$, in $D \cap (K)$, where $(K)$ denotes the disc $|z - z_0| < r$. Let $\varrho$ be a positive number such that $|f(z) - \alpha| > \varrho > 0$ on $L$. We denote by $D_0$ the domain bounded by $L$ and the arc $(e^{i\theta}; \theta_1 \leqq \theta \leqq \theta_2)$. Since $\alpha$ is a cluster value of $f(z)$ at $z_0$, there exists a point $b$ in $D_0$ such that $\beta = f(b)$ lies in $(c)$: $|w - \alpha| < \varrho$. Since $(c) \cap A(f, |\theta - \theta_0| < \eta)$ is of linear measure zero, we may assume that the continuation of rational character of the

---

[1] The following is similar to and much easier than the arguments used in the proof of Theorem 10, § 1.

inverse element $e(w, \beta)$ along the segment joining $\beta$ to $\alpha$ defines a transcendental singularity at $w = \alpha$. Thus it is concluded that $\alpha \in \chi(f, z_0)$.

(III$_2$) Suppose that $\mathscr{C}R(f, z_0)$ contains at most two points. Assume $\alpha \in \mathscr{C}R(f, z_0)$. We select a positive number $r$, sufficiently small, such that $f(z) \neq \alpha$ in $D$ for $|z - z_0| < r$ and $K$: $|z - z_0| = r$ intersects the arc $(e^{i\theta}, |\theta - \theta_0| < \eta)$ at two points. Since $A(f, |\theta - \theta_0| < \eta)$ is of linear measure zero, the set $\Xi$ of arguments $\psi = \arg(w - \alpha)$ for all $w$ belonging to $A(f, |\theta - \theta_0| < \eta) - (\alpha)$ is of linear measure zero. Accordingly, we can find a sequence of points $\{w_n\}$ such that

$$w_n \neq \alpha \text{ for all } n\,, \quad \arg(w_n - \alpha) \notin \Xi\,,$$
$$\arg(w_k - \alpha) \neq \arg(w_h - \alpha) \text{ for } k \neq h \quad \text{and} \quad \lim_{n\to\infty} w_n = \alpha\,.$$

We denote by $g_n$ the segment joining $w_n$ to $\alpha$. If there is another point $\beta \in \mathscr{C}R(f, z_0)$, we may assume that, for all $n$, $g_n$ does not contain $\beta$. Since $w_n \in R(f, z_0)$, there exists a point $z_n$ in $D \cap (K)$ such that $w_n = f(z_n)$, where $(K)$ denotes the disc $|z - z_0| < r$. Consider now the image $\Lambda_n$, in $D \cap (K)$, of the segment $g_n$ by the continuation of the corresponding element $e(w, w_n)$ along $g_n$. Then, either $\Lambda_n$ terminates at a point on $K$ or the end of $\Lambda_n$ lies on $\Gamma$. In case there is an infinite number of $\Lambda_n$ which terminates at a point of $K$, it is clear that $\alpha \in \Phi(f, z_0)$. In the other case, there exists at least one $\Lambda_n$ whose end lies on $\Gamma$; then $\alpha$ must be an asymptotic value along $\Lambda_n$. Thus we conclude that $\alpha \in \chi(f, z_0) \cup \Phi(f, z_0)$.

Remark. In case $C(f)$ is sub-total, the main theorem in the large is an immediate consequence of the main theorem in the small. Let $\alpha \in \mathscr{F}C(f)$. Then, there exists a point $z_0 \in \Gamma$ such that $\alpha \in \mathscr{F}C(f, z_0) \subset \subset \chi^*(f, z_0) \subset \bar{A}(f)$. Accordingly, $C(f) \cap \mathscr{C}\bar{A}(f)$ is an open set. Next, suppose that $\beta \in C(f) \cap \mathscr{C}\bar{A}(f)$. Then, there exists a point $\zeta_0$ of $\Gamma$ such that

$$\beta \in C(f, \zeta_0) \cap \mathscr{C}\chi^*(f, \zeta_0) \subset \mathscr{I}R(f, \zeta_0) \subset \mathscr{I}R(f)\,,$$

by the relation (12). Thus it is proved $\overline{\mathscr{C}R}(f) \cap C(f) \subset \bar{A}(f)$.

Remark. If $\chi(f, z_0)$ is unrestricted, the relation (10) is best possible in the sense that $\chi^*(f, z_0)$ can not be replaced by $\chi(f, z_0)$ (cf. Remark of Theorem 1).

**4. Related theorems.** The main theorem in the small, i. e. Theorem 2 contains the Iversen-Gross theorem which states: *Let $f(z)$ be meromorphic in $|z| < 1$, then any value of $\mathscr{C}R(f, z_0) \cap C(f, z_0) \cap \mathscr{C}C_\Gamma(f, z_0)$ is an asymptotic value of $f(z)$ on some path terminating at $z_0$* (cf. Theorem 1, § 3, II). Furthermore, Theorem 2 gives an extension of the following Iversen-Gross theorem: *If $f(z)$ is meromorphic in $|z| < 1$, then $\mathscr{C}R(f, z_0) \cap C(f, z_0) \cap \mathscr{C}\Psi^*(f, z_0)$ contains at most two values, where $\Psi^*(f, z_0) \equiv \bigcap_\eta \bar{A}(f, 0 < |\theta - \theta_0| < \eta)$. If this set contains two values, then $\mathscr{C}R(f, z_0)$ contains no* *more than these two values* (cf. COLLINGWOOD-CARTWRIGHT [1], p. 136, also Theorem 7 and its related theorems in § 3).

As already stated, Theorem 2 is closely related to a theorem of OHTSUKA [1] and its extension Theorem 9, § 1 which means $\mathscr{F}C(f, z_0) \subset \subset \mathscr{F}C^*_{\Gamma-E}(f, z_0)$. We now show that Theorem 9, § 1 can be applied to prove a theorem of CARATHÉODORY [2]. For that purpose, we need some preliminaries. Let $w = f(z)$ be meromorphic in $|z| < 1$. We make first a normal covering of the Riemann $w$-sphere by open circular domains (caps) $K_\nu$ ($\nu = 1, 2, \ldots$), i. e., a countable set of circular domains with the property that an arbitrary small circular domain $\varkappa$ of the sphere with center $P$ contains at least one circular domain $K_\nu$ containing $P$. Let $E_\nu$ be the set of points $\zeta = e^{i\theta}$ on $\Gamma$: $|z| = 1$ for which either $\lim_{r\to 1} f(re^{i\theta})$ does not exist, or, if it exists, the limit value lies in $K_\nu$. Let $\{\mathfrak{A}_p\}$ be a monotonously decreasing open arcs on $\Gamma$ having a point $z_0$ in common and whose lengths tend to zero. For each $\nu$ we form the sequence of intersections

$$E_\nu \cap \mathfrak{A}_1, \quad E_\nu \cap \mathfrak{A}_2, \ldots, \quad E_\nu \cap \mathfrak{A}_p, \ldots$$

and denote by $n_j$ ($j = 1, 2, \ldots$) that integer (if it exists) for which at least one of the sets $E_{n_j} \subset \mathfrak{A}_p$ ($p = 1, 2, \ldots$) has linear measure zero. After determining the $n_j$'s we form next the open set $G = \bigcup_{j=1}^{\infty} K_{n_j}$ and denote its closed complement by $H$[1]. The set $H$ can never be empty, otherwise $G$ would be the whole $w$-sphere so that a finite number of the $K_{n_j}$'s would cover the sphere. Hence there must exist an arc $\mathfrak{A}_p$ such that for every point $\zeta = e^{i\theta}$ of $\mathfrak{A}_p$ the radial limit either does not exist or is equal to an arbitrary complex number. This arc $\mathfrak{A}_p$ must be a null set which is impossible. Every point $\omega$ of $H$ is a cluster value of $f(z)$ at $z_0$. To prove this, we assume that there is a value $\omega$ of $H$ not in $C(f, z_0)$; then there is a neighborhood $U(z_0)$ of $z_0$ such that $|f(z) - \omega| \geqq \varrho > 0$ in the intersection of $U(z_0)$ with $|z| < 1$. There exists a $K_\nu$ containing $\omega$ and contained in $|w - \omega| < \varrho$. By Fatou's theorem, the radial limit $f(e^{i\theta}) = \lim_{r\to 1} f(re^{i\theta})$ exists for almost all $e^{i\theta}$ of $\mathfrak{A}_p$ where $\mathfrak{A}_p \subset U(z_0)$. These limit values $f(e^{i\theta})$ satisfy the inequality $|f(e^{i\theta}) - \omega| \geqq \varrho$ so that $f(e^{i\theta}) \notin K_\nu$. From the definition, it follows that $\nu$ is one of $n_j$'s above, hence $K_\nu \subset G$, so that $\omega \notin H$, which is a contradiction.

**Theorem of CARATHÉODORY** [2]. *Let $w = f(z)$ be meromorphic in $D$: $|z| < 1$ and let $z_0$ be any point of $|z| = 1$. Let $G$ be decomposed into its*

[1] $G$ can be empty as is shown by the case of the elliptic modular function, the function which maps the $w$-plane punctured at 0, 1, $\infty$ onto the unit disk conformally. The radial limit of the inverse function $w = f(z)$ can exist when this limit is 0, 1, or $\infty$, and this case occurs only when $\zeta = e^{i\theta}$ is one of the cusps of the modular figure. In this case the linear measure of $E_\nu \cap \mathfrak{A}_p$ is positive for every $\nu$ and every $p$; hence the set $G$ is empty.

*connected components,* $G = \bigcup_n G_n$. *Then either* $G_n$ *contains no point of* $C(f, z_0)$ *or else is contained in* $C(f, z_0)$. *Hence,* $C(f, z_0)$ *is the union of* $H$ *and a certain number of* $G_n$.

*Proof.* Let $\alpha \in G_n$ not be a point of $C(f, z_0)$. It is sufficient to show that $\bar{\Delta}_w \cap C(f, z_0) = \theta$ for any closed Jordan domain $\bar{\Delta}_w$ containing $\alpha$ and contained in $G_n$. First we cover $\bar{\Delta}_w$ by the union $\mathfrak{D}$ of a finite number of suitably chosen $K_{n_j}$. Since $\alpha \notin C(f, z_0)$, the function $\varphi(z) = 1/(f(z) - \alpha)$ is regular and bounded in a certain neighborhood of $z_0$, so that Fatou's theorem holds on some arc $\mathfrak{A}_q$. Therefore $f(z)$ has radial limits $f(e^{i\theta}) = \lim_{r \to 1} f(re^{i\theta})$ almost everywhere on $\mathfrak{A}_{p_0}$. If we take $p_0$ sufficiently large, the set $E$ of points $\zeta = e^{i\theta}$ of $\mathfrak{A}_{p_0}$, for which the radial limit $f(e^{i\theta})$ does not exist or $f(e^{i\theta})$ belongs to $\mathfrak{D}$, has measure zero. Using the radial limit $f(e^{i\theta})$ for every $e^{i\theta} \in \mathfrak{A}_{p_0} - E - z_0$, we can define the boundary cluster set $C^*_{\Gamma - E}(f, z_0)$. Clearly $C^*_{\Gamma - E}(f, z_0)$ has no points in common with $\mathfrak{D}$, hence $C^*_{\Gamma - E}(f, z_0) \cap \bar{\Delta}_w = \theta$. Accordingly $C(f, z_0) \cap \bar{\Delta}_w = \theta$.

Remark. The following example illustrates the meaning of Carathéodory's theorem. Let $w = f(z)$ be the function which maps conformally the circular triangle $\Delta_z$ in $D$: $|z| < 1$ with vertices at $z = 1$, $e^{\frac{2\pi}{3}i}$, $e^{\frac{4\pi}{3}i}$ and sides orthogonal to $|z| = 1$ onto a circular triangle $\Delta_w$ in $|w| < 1$ with angles $\pi/2$, $\pi/4$, $\pi/6$ and sides orthogonal to $|w| = 1$ so that vertices of $\Delta_z$ correspond to those of $\Delta_w$. As in the case of the modular function, we continue $w = f(z)$ by reflection and denote also by $w = f(z)$ the resulting function. The function $w = f(z)$ is a function of class $(U)$ in Seidel's sense and has every point of $|z| = 1$ as an essential singularity. The set of points $\zeta = e^{i\theta}$ where the radial limits $f(e^{i\theta})$ are vertices of $\Delta_w$ or vertices of successive reflections of $\Delta_w$ is countable. The cluster set $C(f, z_0)$ at every point $z_0$ of $\Gamma$: $|z| = 1$ consists of the closed disc $|w| \leqq 1$. Let $\mathfrak{A}_w$ be any arbitrarily small arc of $|w| = 1$. For any open circular arc $\mathfrak{A}_z$ containing $z_0$ on $|z| = 1$, the set of points $e^{i\theta}$ of $\mathfrak{A}_z$ where the limit values $f(e^{i\theta})$ of $w = f(z)$ lie in $\mathfrak{A}_w$ has positive measure; otherwise the set $H$ would contain no point of $\mathfrak{A}_w$ so that $C(f, z_0)$ coincides with the whole $w$-plane, which is a contradiction.

## § 3. Baire category and cluster sets

1. Bagemihl-Seidel [1] and Collingwood [3] have obtained independently important results on cluster sets by using the notion of Baire category.

Bagemihl-Seidel's results are based on the following general principle.

**Theorem 1.** *Let* $\Omega$ *be a non-empty complete metric space. Let the class* $\{\mathscr{P}_n\}$ *consist of a finite or an enumerable number of properties, each of which is meaningful for every element* $\omega \in \Omega$. *Suppose that the following conditions are satisfied:*

(1) *For every* $\mathscr{P}_n$, *if* $X$ *is a dense subset of some non-empty open set* $G \subset \Omega$, *and if every* $x \in X$ *has property* $\mathscr{P}_n$, *then every* $g \in G$ *has property* $\mathscr{P}_n$.

(2) *For every* $\mathscr{P}_n$, *every non-empty open subset of* $\Omega$ *contains at least one element which has property* $\mathscr{P}'_n$, *where* $\mathscr{P}'_n$ *denotes the negation of* $\mathscr{P}_n$.

*Then, there exists a residual set*[1] $S \subset \Omega$, *every element of which has property* $\mathscr{P}'_n$ *for all* $n$ (BAGEMIHL-SEIDEL [1]).

*Proof.* Let $E_n$ be the set of points of $\Omega$, each of which possesses property $\mathscr{P}_n$. Then $E_n$ is nowhere dense in $\Omega$; otherwise $E_n$ would be dense in some non-empty open set, $G$, of $\Omega$, and every element of $G$, by (1), would have property $\mathscr{P}_n$, contradicting (2).

As its application, we prove

**Theorem 2.** *Let* $\mathscr{H}_1$ *be a Hausdorff space,* $\mathscr{H}_2$ *a Hausdorff space satisfying the second axiom of countability, and* $\Omega$ *a complete metric space, all non-empty. To every element* $\omega \in \Omega$, *let there correspond a non-empty subset* $H_\omega \subset \mathscr{H}_1$ *such that*

(3) *if a set* $X \subset \Omega$ *is dense in some non-empty open set* $G \subset \Omega$, *then* $H_X$ *is dense in* $H_G$[2].

*Let* $f$ *be a continuous mapping* $\mathscr{H}_1$ *into* $\mathscr{H}_2$, *such that*

(4) *if* $G$ *is a non-empty open subset of* $\Omega$, *then* $f(H_G)$ *is dense in* $\mathscr{H}_2$.

*Then there exists a residual set* $S \subset \Omega$ *such that, for every* $s \in S$, $f(H_s)$ *is dense in* $\mathscr{H}_2$ (BAGEMIHL-SEIDEL [1]).

*Proof.* Let $V_n$ be a basis of $\mathscr{H}_2$. For every $\omega \in \Omega$, we let $\mathscr{P}_n$ be the property that $f(H_\omega) \cap V_n = \emptyset$. Note that (1) follows from the continuity of $f$ and (3), and that (4) implies (2). Then, Theorem 2 is an immediate consequence of Theorem 1.

Now we state some applications of Theorem 2 to the theory of cluster sets.

Let $f(z)$ be meromorphic in $D$: $|z| < 1$. Suppose that $f(z)$ maps $D$ on a Riemann surface $\mathscr{R}$. We define the cluster set $C_D(f, e^{i\theta})$, relative to $\mathscr{R}$, of $f(z)$ at $z = e^{i\theta}$ to be the set of all points $w$ of $\mathscr{R}$ with the property that there exists a sequence $\{z_n\}$ of points in $D$, such that $z_n \to e^{i\theta}$ and, on $\mathscr{R}$, $f(z_n) \to w$. Similarly, we can define the radial cluster set $C_\varrho(f, e^{i\theta})$, relative to $\mathscr{R}$, of $f(z)$ at $z = e^{i\theta}$.

**Theorem 3.** *Let* $f(z)$ *be meromorphic in* $D$: $|z| < 1$ *and map* $|z| < 1$ *on the Riemann surface* $\mathscr{R}$. *Suppose that the cluster set* $C_D(f, e^{i\theta})$, *relative to* $\mathscr{R}$, *of* $f(z)$ *at every point of an arc* $\mathfrak{A}$ $(e^{i\theta};\ \theta_1 \leqq \theta \leqq \theta_2,\ \theta_1 < \theta_2)$ *is* $\mathscr{R}$. *Then there exists a residual set* $S \subset \mathfrak{A}$ *such that the radial cluster set* $C_\varrho(f, e^{i\theta})$, *relative to* $\mathscr{R}$, *of* $f(z)$, *corresponding to every* $s \in S$, *is* $\mathscr{R}$ (BAGEMIHL-SEIDEL [1]).

---

[1] A residual set is the complement of a set of category I.

[2] $H_X = \bigcup_{x \in X} H_x$.

*Proof.* This theorem is an immediate consequence of Theorem 2. We have only to identify $\mathscr{H}_1$ with $D$, $\mathscr{H}_2$ with $\mathscr{R}$, $\Omega$ with $\mathfrak{A}$, $H_\omega$, for every $\omega = e^{i\theta} \in \Omega$, with the radius $\varrho$ terminating at $e^{i\theta}$, and $f$ with $f(z)$.

Using usual definitions of $C_D(f, e^{i\theta})$ and $C_\varrho(f, e^{i\theta})$, we have the following special case of Theorem 3.

**Theorem 4.** *Let $f(z)$ be meromorphic in $D$: $|z| < 1$ and let its domain of values be the domain $\mathfrak{D}$ in the complex $w$-plane. Suppose that the cluster set $C_D(f, e^{i\theta})$ of $f(z)$ at every point of an arc $\mathfrak{A}$ is $\overline{\mathfrak{D}}$. Then there exists a residual set $S$ such that the radial cluster set $C_\varrho(f, e^{i\theta})$ of $f(z)$ corresponding to every $e^{i\theta} \in S$ is $\overline{\mathfrak{D}}$* (BAGEMIHL-SEIDEL [1])[1].

To apply Theorem 4 for the case in which $\mathfrak{D}$ is the complex $w$-plane, we introduce some definitions. Let $f(z)$ be meromorphic in $D$: $|z| < 1$. Let $\mathscr{M}$ be a subset of an arc $\mathfrak{A}$. If the intersection of $\mathscr{M}$ with every subarc of $\mathfrak{A}$ is of positive measure, then $\mathscr{M}$ is said to be *metrically dense* on $\mathfrak{A}$. Further, if $\lim f(z)$ exists uniformly as $z \to e^{i\theta}$ in every Stolz angle with vertex $e^{i\theta}$, then the point $e^{i\theta}$ is called a Fatou point of $f(z)$.

**Theorem 5.** *Let $f(z)$ be meromorphic in $|z| < 1$. If any one of the following conditions is satisfied:*

(i) *$f(z)$ has no limit along any radius terminating at a point of $\mathscr{M}$, where $\mathscr{M}$ is metrically dense on an arc $\mathfrak{A}$; or*

(ii) *$f(z)$ is non-constant in $|z| < 1$ and the radial cluster set $C_\varrho(f, e^{i\theta})$ contains a fixed constant $\alpha$, finite or infinite, for every $e^{i\theta}$ of $\mathscr{M}$ metrically dense on $\mathfrak{A}$; or*

(iii) *$\mathscr{M}$ is the complement of the set of Fatou points of $f(z)$ and metrically dense on $\mathfrak{A}$; then there exists a residual set $S \subset \mathfrak{A}$ such that the radial cluster set $C_\varrho(f, e^{i\theta})$ corresponding to every point $e^{i\theta} \in S$ is the whole $w$-plane, i. e., total* (BAGEMIHL-SEIDEL [1]).

*Proof.* In virtue of Theorem 4, it is sufficient to show that $C_D(f, e^{i\theta})$ is total for every $e^{i\theta}$ of the arc $\mathfrak{A}$. Under hypothesis (iii), this follows immediately from Plessner's theorem[2] which states: If $f(z)$ is meromorphic in $|z| < 1$, then almost every point $e^{i\theta}$ of $|z| = 1$ is either a Fatou point or a Plessner point[3]. Under hypothesis (i) or (ii), let us suppose that, on the contrary, the cluster set $C_D(f, z_0)$ at some point $z_0$ of $\mathfrak{A}$ is not total. Let $\beta \notin C_D(f, z_0)$. Then $g(z) = 1/(f(z) - \beta)$ is bounded in some neighborhood $U(z_0)$ of $z_0$. By an extension of Fatou's theorem, $f(z)$ has radial limit $f(e^{i\theta})$ for almost every $e^{i\theta}$ of $\mathscr{M} \cap U(z_0)$. This already contradicts (i). In case (ii), $g(z)$ has the same radial limit $1/(\alpha - \beta)$ for a

---

[1] In case $\mathfrak{D}$ is the unit circle: $|w| < 1$, this theorem, e. g., applies to any Blaschke product $f(z)$ whose zeros have every point of the arc $\mathfrak{A}$ as an accumulation point (SEIDEL [2], p. 211).

[2] PLESSNER [1]. See § 4.

[3] The point $e^{i\theta}$ is called a Plessner point if the cluster set $C_\Delta(f, e^{i\theta})$ is total for any Stolz angle $\Delta$ with vertex $e^{i\theta}$.

set of points $e^{i\theta}$ of positive measure on $\mathcal{M} \cap U(z_0)$. By Riesz' theorem, it follows that $g(z)$, and hence also $f(z)$ must be identically constant, contrary to hypothesis.

**Theorem of Lusin and Privaloff**[1]. *If $f(z)$ is regular in $|z| < 1$ and has the same radial limit $\alpha$ for a set $\mathcal{M}_\alpha(\theta)$ of points $z = e^{i\theta}$, $\mathcal{M}_\alpha(\theta)$ being both metrically dense and of category II on an arc $\mathfrak{A}$ of $|z| = 1$, then $f(z) \equiv \alpha$.*

As an immediate consequence, Theorem 4 (ii) gives a generalization of the above uniqueness theorem:

**Corollary 1.** *Let $f(z)$ be meromorphic in $|z| < 1$. Let $\mathcal{M}$ be metrically dense on an arc $\mathfrak{A}$ and a residual subset of $\mathfrak{A}$. Suppose that for every $e^{i\theta} \in \mathcal{M}$ the radial cluster set $C_\varrho(f, e^{i\theta})$ contains a fixed value $\alpha$ but is not total. Then $f(z) \equiv \alpha$.*

Remark. Theorem 4 (ii) applies, for example, to the case of a non-constant meromorphic function $f(z)$ which converges uniformly to a fixed constant $\alpha$ (finite or infinite) on a sequence of Jordan closed curves each of which contains the preceding one in its interior, or to a non-constant meromorphic function $f(z)$ which converges to $\alpha$ on a spiral in $|z| < 1$ whose end is the whole circumference $|z| = 1$[2].

Remark. It is proved that given a number $\lambda$, $0 \leqq \lambda \leqq 2\pi$, there exists a meromorphic function $f(z)$ in $|z| < 1$ such that the set $S$ consisting of all those points $e^{i\theta}$, for each of which the corresponding radial cluster set of $f(z)$ is total, is of measure $\lambda$; in the case $\lambda = 0$, the set $S$ may be finite, countable or of the power of the continuum[3].

2. F. Wolf [1] has proved a theorem related to the Lusin-Privaloff theorem.

**Wolf's theorem.** *Suppose that $f(z)$ is regular in $|z| < 1$ and that there is a set $\mathcal{M}(\theta)$ of points $z = e^{i\theta}$, $\mathcal{M}(\theta)$ being a set of $G_\delta$ type dense on an arc $\mathfrak{A}$ of $|z| = 1$, such that $\infty \notin C_\varrho(f, e^{i\theta})$ for all $e^{i\theta} \in \mathcal{M}(\theta)$. Then, if there is a number $\alpha \neq \infty$ such that $\alpha \in C_\varrho(f, e^{i\theta})$ for almost every $e^{i\theta} \in \mathfrak{A}$, we have $f(z) \equiv \alpha$.*

The starting point of Collingwood's investigation [3] is to obtain a theorem (see Theorem 6) which contains both the Lusin-Privaloff theorem and the Wolf theorem, by combining the ideas in the proofs of them.

To state Collingwood's results, we need some preliminaries. Let $w = f(z)$ be meromorphic in the unit circle $D$: $|z| < 1$. In general, we define the cluster set $C_S(f, e^{i\theta})$, on a set $S (\subset D)$, of $f(z)$ at $z = e^{i\theta}$, to be the set of all values $\alpha$ with the property that there exists a sequence $\{z_n\}$ of points on $S$ such that $z_n \to e^{i\theta}$ and $f(z_n) \to \alpha$. If we adopt as $S$ the chord $\varrho(\varphi)$ of the unit circle through $e^{i\theta}$ and inclined at the angle

[1] Lusin and Privaloff [1].

[2] Valiron [1], Bagemihl-Erdös-Seidel [1]; for the related results, see § 4.

[3] Bagemihl and Seidel [1], p. 1072.

$\varphi, -\frac{\pi}{2} < \varphi < \frac{\pi}{2}$, to the radius through $e^{i\theta}$, positive angles being measured to the right of the radius and negative angles to the left, we can define the chordal cluster set $C_{\varrho(\varphi)}(f, e^{i\theta})$; clearly $C_{\varrho(0)}(f, e^{i\theta})$ coincides with the radial cluster set $C_{\varrho}(f, e^{i\theta})$. Similarly, we can define the angular cluster set $C_{\Delta}(f, e^{i\theta})$ for any Stolz angle $\Delta$ with vertex $e^{i\theta}$; $C_{\Delta}(f, e^{i\theta})$ is either a single point or a continuum. A cluster set or union of cluster sets is said to be degenerate if it consists of a single point; otherwise it is non-degenerate. A cluster set or union of cluster sets whose complement is empty, we call *total*; one whose complement is not empty we call *subtotal*, the degenerate cluster sets being a sub-class of the subtotal cluster sets. As already stated, if $\bigcup_{\Delta} C_{\Delta}(f, e^{i\theta})$ taken over all angles $\Delta$ between pairs of chords through $z = e^{i\theta}$ is degenerate, $e^{i\theta}$ is a Fatou point and if $C_{\Delta}(f, e^{i\theta})$ is total for any Stolz angle $\Delta$ with vertex $e^{i\theta}$, then $e^{i\theta}$ is a Plessner point. The sets of Fatou points and Plessner points are denoted by $F(f)$ and $I(f)$ respectively. We define a Fatou *arc* as an arc of $|z| = 1$ which is an open arc of the frontier of a simply-connected Jordan domain $G$ in $|z| < 1$ in which either $f(z)$ or for some $\beta \neq \infty$, $1/(f(z) - \beta)$ is bounded. From Fatou's theorem follows that almost every point of a Fatou arc belongs to $F(f)$.

**Lemma 1.** *Suppose that $w = f(z)$ is meromorphic in $|z| < 1$ and that for some fixed $\varphi$, $-\frac{\pi}{2} < \varphi < \frac{\pi}{2}$, and some complex number $\beta$, finite or infinite, there is a set $\mathcal{M}(\theta)$ of points $z = e^{i\theta}$ of category II on a certain arc $\mathfrak{A}$ of $|z| = 1$ and such that $\beta \notin C_{\varrho(\varphi)}(f, e^{i\theta})$ for all $e^{i\theta} \in \mathcal{M}(\theta)$. Then the arc $\mathfrak{A}$ contains an arc $\mathfrak{A}_0$ such that*

(i) *$\mathfrak{A}_0$ is a Fatou arc for $f(z)$ in the neighborhood of which either $f(z)$ or $1/(f(z) - \beta)$ is uniformly bounded according as $\beta = \infty$ or $\beta \neq \infty$, and*

(ii) *$\mathcal{M}(\theta)$ is dense on $\mathfrak{A}_0$* (Collingwood [3]).

*Proof.* Without loss of generality we may suppose that $\beta = \infty$. Further, we assume that $\mathcal{M}(\theta) \subset \mathfrak{A}$. Denote by $E_{\varphi}(\sigma, N, \theta)$ the set of points $e^{i\theta} \in \mathcal{M}(\theta)$, such that, for all values $t$ in $0 < t < \sigma$

$$|f(e^{i\theta}(1 - te^{i\varphi}))| < N .$$

Now take

$$\sigma_1 > \sigma_2 > \cdots > \sigma_\nu > \cdots, \quad \lim_{\nu \to \infty} \sigma_\nu = 0 .$$

Then

$$E_{\varphi}(\sigma_\nu, N, \theta) \subset E_{\varphi}(\sigma_{\nu+1}, N, \theta) .$$

Also, for any $\eta > 0$,

$$E_{\varphi}(\sigma, N, \theta) \subset E_{\varphi}(\sigma, N + \eta, \theta)$$

so that if we take

$$N_1 < N_2 < \cdots < N_\nu < \cdots ,$$

then

$$E_\nu = E_\varphi(\sigma_\nu, N_\nu, \theta) \subset E_\varphi(\sigma_{\nu+1}, N_{\nu+1}, \theta) = E_{\nu+1}\,.$$

Obviously

$$\mathcal{M}(\theta) = \bigcup_\nu E_\nu\,. \tag{5}$$

By hypothesis, since $\mathcal{M}(\theta)$ is of category II on $\mathfrak{A}$ it follows that at least one of the sets $E_\nu$, say $E_k$, is of category II on $\mathfrak{A}$. Therefore, there is a subarc $\mathfrak{A}_0$ of $\mathfrak{A}$ such that $E_k$ is dense on $\mathfrak{A}_0$; since $E_k \subset \mathcal{M}(\theta)$, the set $\mathcal{M}(\theta)$ is dense on $\mathfrak{A}_0$. For $e^{i\theta} \in E_k \cap \mathfrak{A}_0$ and for all $0 < t < \sigma_k$

$$|f(e^{i\theta}(1 - te^{i\varphi}))| < N_k\,; \tag{6}$$

and since $E_k$ is dense on $\mathfrak{A}_0$, it follows that the inequality

$$|f(z)| \leqq N_k \tag{7}$$

is satisfied throughout the annular quadrilateral $G$, not containing the origin, defined by the arc $\mathfrak{A}_0$, the two chords at the end points of $\mathfrak{A}_0$ inclined at the angle $\varphi$ to the respective radii at these points, and the circular arc $|z| = r$ passing through the end points of the chords. Because, every point of $G$ is, by (6) and the fact that $E_k$ is dense on $\mathfrak{A}_0$, an accumulation point of points at which $|f(z)| < N_k$, so that (7) is satisfied at every point of $G$, since $f(z)$ is meromorphic in $|z| < 1$. This completes the proof of the lemma.

As a first application of Lemma 1, we have the following general theorem.

**Theorem 6.** *If $f(z)$ is meromorphic in $|z| < 1$ and if, for a constant $\varphi$, $-\frac{\pi}{2} < \varphi < \frac{\pi}{2}$, there is a set $\mathcal{M}(\theta)$ of points $z = e^{i\theta}$ of category II on an arc $\mathfrak{A}$ of $|z| = 1$ such that*

$$\bigcap_{e^{i\theta} \in \mathcal{M}(\theta)} \mathscr{C} C_{\varrho(\varphi)}(f, e^{i\theta}) \neq \emptyset$$

*and if, further, there is a number $\alpha$, finite or infinite, and a set $\mathcal{N}(\theta)$ metrically dense on $\mathfrak{A}$ such that*

$$\alpha \in \bigcap_{e^{i\theta} \in \mathcal{N}(\theta)} C_{\varrho(\varphi)}(f, e^{i\theta})\,,$$

*then $f(z) \equiv \alpha$* (Collingwood [3]).

*Proof.* By Lemma 1, $\mathfrak{A}$ contains an arc $\mathfrak{A}_0$ which is a Fatou arc for $f(z)$. By Fatou's theorem, $F(f)$ is almost everywhere on $\mathfrak{A}_0$ and so, since $m(\mathcal{N}(\theta) \cap \mathfrak{A}_0) > 0$ by hypothesis, it follows that $m(\mathcal{N}(\theta) \cap \mathfrak{A}_0 \cap F(f)) > 0$. For $e^{i\theta} \in F(f)$ we have an angular limit $f(e^{i\theta})$ and so $C_\varrho(f, e^{i\theta}) = \alpha$ for every $e^{i\theta} \in \mathcal{N}(\theta) \cap \mathfrak{A}_0 \cap F(f)$. Accordingly, by Riesz' theorem, $f(z) \equiv \alpha$.

To supplement Lemma 1, we state

**Lemma 2.** *If $w = f(z)$ is meromorphic in $|z| < 1$ and if, for some fixed $\varphi$, $-\frac{\pi}{2} < \varphi < \frac{\pi}{2}$, there is a set $\mathcal{M}(\theta)$ of category II on an arc $\mathfrak{A}$*

*of the circumference* $|z| = 1$ *such that* $C_{\varrho(\varphi)}(f, e^{i\theta})$ *is sub-total for all* $e^{i\theta} \in \in \mathcal{M}(\theta)$, *then there is a subset* $\mathcal{M}_0(\theta)$ *of* $\mathcal{M}(\theta)$, *also of category II on* $\mathfrak{A}$, *such that*

$$\bigcap_{\mathcal{M}_0} \mathscr{C} C_{\varrho(\varphi)}(f, e^{i\theta}) \neq \emptyset$$

(Collingwood [3]).

*Proof.* If $C_{\varrho(\varphi)}(f, e^{i\theta})$ is sub-total, we can find a circular disc contained in $\mathscr{C} C_{\varrho(\varphi)}(f, e^{i\theta})$, since this is an open set. Now suppose the $w$-plane to be projected stereographically onto the Riemann $w$-sphere. We construct on the $w$-sphere a sequence of finite triangular lattices $l_1, l_2, \ldots, l_n, \ldots$, each lattice being a sub-division of its predecessor, such that all the diameters of triangles of the lattice $l_n$ are less than $1/2^n$. We denote the individual triangles in the lattice $l_n$ by $\Delta_{n,1}, \Delta_{n,2}, \ldots, \Delta_{n,m(n)}$. Denote by $\Gamma_n(\theta)$ the set of points $e^{i\theta} \in \mathcal{M}(\theta)$ such that $n$ is the smallest number for which $\mathscr{C} C_{\varrho(\varphi)}(f, e^{i\theta})$ contains completely at least one of the triangles $\Delta_{n,\nu}$ $(1 \leqq \nu \leqq m(n))$. Then, clearly $\Gamma_1(\theta) \subset \Gamma_2(\theta) \subset \cdots \subset \Gamma_n(\theta) \subset \cdots$ and $\mathcal{M}(\theta) = \bigcup_{n=1}^{\infty} \Gamma_n(\theta)$. We now sub-divide the set $\Gamma_n(\theta)$ in the following way: We assign to each triangle $\Delta_{n,1}, \Delta_{n,2}, \ldots, \Delta_{n,m(n)}$ all those values of $e^{i\theta} \in \Gamma_n(\theta)$ for which $\Delta_{n,\nu}$, $1 \leqq \nu \leqq m(n)$, is contained in $\mathscr{C} C_{\varrho(\varphi)}(f, e^{i\theta})$, and we denote the corresponding sub-sets of $\Gamma_n(\theta)$ by $E_{n,\nu}(\theta)$, $1 \leqq \nu \leqq \leqq m(n)$. Evidently, at least one of $E_{n,\nu}(\theta)$ is non-empty and two non-empty $E_{n,\nu}(\theta)$ may have common points. We have $\mathcal{M}(\theta) = \bigcup_{n} \bigcup_{1 \leqq \nu \leqq m(n)} E_{n,\nu}(\theta)$. Since $\mathcal{M}(\theta)$ is of category II on $\mathfrak{A}$, there exists at least one $E_{n,\nu}(\theta)$, $n < \infty$, $\nu \leqq m(n)$, say, $E_{j,k}(\theta)$ of category II on $\mathfrak{A}$. However, for all $e^{i\theta} \in E_{j,k}(\theta)$, the triangle $\Delta_{j,k}$ is contained in $\mathscr{C} C_{\varrho(\varphi)}(f, e^{i\theta})$ and so, by putting $\mathcal{M}_0(\theta) = E_{j,k}(\theta)$, the lemma is proved.

With the aid of Lemma 2, we can easily sharpen Theorem 6 in the following form.

**Theorem 7.** *If* $w = f(z)$ *is meromorphic in* $|z| < 1$ *and if, for a constant* $\varphi$, $-\frac{\pi}{2} < \varphi < \frac{\pi}{2}$, *there is a set* $\mathcal{M}(\theta)$ *of points* $z = e^{i\theta}$ *of category II on an arc* $\mathfrak{A}$ *of* $|z| = 1$ *such that* $C_{\varrho(\varphi)}(f, e^{i\theta})$ *is sub-total for all* $e^{i\theta} \in \mathcal{M}(\theta)$ *and if, further, there is a number* $\alpha$, *finite or infinite, and a set* $\mathcal{N}(\theta)$ *metrically dense on* $\mathfrak{A}$ *such that*

$$\alpha \in \bigcap_{e^{i\theta} \in \mathcal{N}(\theta)} C_{\varrho(\varphi)}(f, e^{i\theta}),$$

*then* $f(z) \equiv \alpha$.

By Theorem 7, we can improve Corollary 1.

**Corollary 2.** *Let* $f(z)$ *be meromorphic in* $|z| < 1$. *Let* $\mathcal{M}(\theta)$ *be metrically dense and of category II on an arc* $\mathfrak{A}$ *of* $|z| = 1$. *Suppose that for every* $e^{i\theta} \in \mathcal{M}(\theta)$ *the radial cluster set* $C_{\varrho}(f, e^{i\theta})$ *contains a fixed value* $\alpha$ *but is sub-total. Then* $f(z) \equiv \alpha$.

**Corollary 3.** *If $f(z)$ is meromorphic in $|z| < 1$ and if there is a set $\mathcal{N}(\theta)$ of points $z = e^{i\theta}$ metrically dense on an arc $\mathfrak{A}$ of $|z| = 1$ such that $\bigcap_{\mathcal{N}(\theta)} C_\varrho(f, e^{i\theta}) \neq \emptyset$, then either given any set $\mathcal{M}(\theta)$ of category II on $\mathfrak{A}$, the union $\bigcup_{\mathcal{M}(\theta)} C_\varrho(f, e^{i\theta})$ is total or $f(z)$ is a constant.*

*Proof.* This is an immediate consequence of Theorem 6. Suppose that $\bigcup_{\mathcal{M}(\theta)} C_\varrho(f, e^{i\theta})$ is sub-total so that its complement $\bigcap_{\mathcal{M}(\theta)} \mathscr{C} C_\varrho(f, e^{i\theta}) \neq \emptyset$. If $\alpha \in \bigcap_{\mathcal{N}(\theta)} C_\varrho(f, e^{i\theta})$, then it follows that $f(z) \equiv \alpha$, by Theorem 6.

**Corollary 4.** *If $f(z)$ is non-constant and meromorphic in $|z| < 1$ and if $f(z)$ has the radial limit $f(e^{i\theta}) = \alpha$ for all $e^{i\theta} \in \mathcal{N}(\theta)$ where $\mathcal{N}(\theta)$ is metrically dense on an arc $\mathfrak{A}$ of $|z| = 1$, then*

i) *$\mathscr{C}\mathcal{N}(\theta)$ is residual on $\mathfrak{A}$*

*and*

ii) *$\bigcup_{\mathscr{C}\mathcal{N}(\theta)} C_\varrho(f, e^{i\theta})$ is total.*

*Proof.* We prove that $\mathcal{N}(\theta)$ is of category I. Otherwise, $\mathcal{N}(\theta)$ would be of category II and $f(z)$ would be constant, by Corollary 2. Since $\mathscr{C}\mathcal{N}(\theta)$ is residual and so of category II on $\mathfrak{A}$, we have ii) by Corollary 3.

**3.** To state a property of the set $I(f)$ of Plessner points, we make use of the following

**Lemma 3.** *If $f(z)$ is meromorphic in $|z| < 1$ and if there is a set $\mathcal{M}(\theta)$ of points $z = e^{i\theta}$ satisfying the conditions:*

(a) $\mathcal{M}(\theta) \subset \mathscr{C} I(f)$

*and*

(b) *$\mathcal{M}(\theta)$ is of category II on an arc $\mathfrak{A}$ of the circumference $|z| = 1$, then there is an arc $\mathfrak{A}_0 \subset \mathfrak{A}$ such that*

(i) *$\mathfrak{A}_0$ is a Fatou arc for $f(z)$,*

*and*

(ii) *$\mathcal{M}(\theta)$ is dense on $\mathfrak{A}_0$*

(Collingwood [3]).

*Proof.* Note that every point $e^{i\theta} \in \mathscr{C} I(f)$ is the vertex of an angle $\Delta(\theta)$ in $|z| < 1$ for which the angular cluster set $C_{\Delta(\theta)}(f, e^{i\theta})$ is sub-total. For each natural number $n$, we define a subset $E_n$ of $\mathcal{M}(\theta)$ as follows: $e^{i\theta} \in E_n$, if there is a $\Delta(\theta)$ with vertex $e^{i\theta}$ of magnitude greater than $\pi/2^n$ in which $C_{\Delta(\theta)}(f, e^{i\theta})$ is sub-total. Evidently, $E_1 \subset E_2 \subset \subset \cdots \subset E_n \subset \cdots$ and $\mathcal{M}(\theta) = \bigcup_n E_n$. Now, we sub-divide each set $E_n$ into a finite number of subsets in the following way. We first divide the Stolz angle $-\frac{\pi}{2}(1 - 2^{-n}) < \varphi < \frac{\pi}{2}(1 - 2^{-n})$ with vertex $e^{i\theta}$ into $N = 2^n - 1$ equal parts of magnitude $\pi/2^n$, by drawing $N - 1$ chords through $e^{i\theta}$. We denote these chords (inclusive two sides of the original angle) by $\varrho_1 = \varrho(\varphi_1)$, $\varrho_2 = \varrho(\varphi_2)$, $\ldots$, $\varrho_{N+1} = \varrho(\varphi_{N+1})$. Evidently, each angle $\Delta(\theta)$ of magnitude $> \pi/2^n$ contains at least one of $\varrho_1, \varrho_2, \ldots, \varrho_{N+1}$.

We denote by $E_{n,\nu}$, $1 \leqq \nu \leqq N+1$, the subset of $E_n$ for which there is a $\Delta(\theta)$ of magnitude greater than $\pi/2^n$ such that $C_{\Delta(\theta)}(f, e^{i\theta})$ is sub-total and which contains $\varrho_\nu$. Clearly, a point $e^{i\theta}$ may belong to more than one of the sets $E_{n,\nu}$. Then $\mathcal{M}(\theta) = \bigcup_n \bigcup_{1\leqq\nu\leqq N+1} E_{n,\nu}$. Since $\mathcal{M}(\theta)$ is of category II on $\mathfrak{A}$, at least one of the sets $E_{n,\nu}$, say $E_{j,k}$, must be of category II on $\mathfrak{A}$. Consequently, for all $e^{i\theta} \in E_{j,k}$, $C_{\varrho_k}(f, e^{i\theta})$ is sub-total; and it follows from Lemma 2, that there is a subset $\mathcal{M}_0(\theta) \subset E_{j,k} \subset \mathcal{M}(\theta)$, also of category II on $\mathfrak{A}$ and $\bigcap_{\mathcal{M}_0} \mathscr{C} C_{\varrho_k}(f, e^{i\theta})$ is not empty. Then our assertion follows from Lemma 1.

We prove

**Theorem 8.** *If $f(z)$ is meromorphic in $|z| < 1$ and the set $I(f)$ of Plessner points of $f(z)$ is dense on an arc $\mathfrak{A}$ of the unit circumference $|z| = 1$, then $I(f)$ is also residual on $\mathfrak{A}$* (Collingwood [3]).

*Proof.* Suppose that $\mathscr{C}I(f)$ is of category II on $\mathfrak{A}$. Then, putting $\mathcal{M}(\theta) = \mathscr{C}I(f)$ and applying Lemma 3, we see that there is a Fatou arc $\mathfrak{A}_0 \subset \mathfrak{A}$. Since $\mathfrak{A}_0 \subset \mathscr{C}I(f)$, this implies that $I(f)$ is not dense on $\mathfrak{A}$.

Let $f(z)$ be meromorphic in $|z| < 1$. We define the set $W_\varrho(f)$ as the set of points $z = e^{i\theta}$ for each of which the radial cluster set $C_\varrho(f, e^{i\theta})$ is total. Between $I(f)$ and $W_\varrho(f)$, there is an important relation.

**Theorem 9.** *If $f(z)$ is meromorphic in $|z| < 1$, then the sets $W_\varrho(f)$ and $I(f)$ differ by a set of category I on $|z| = 1$* (Collingwood [3]).

*Proof.* 1°. $W_\varrho(f) \cap \mathscr{C}I(f)$ is of category I. To prove this, put

$$\mathcal{M}(\theta) = W_\varrho(f) \cap \mathscr{C}I(f)\,.$$

Contrary to the assertion, if $\mathcal{M}(\theta)$ is of category II, then by Lemma 3, there exists a Fatou arc $\mathfrak{A}_0$ on which $\mathcal{M}(\theta)$ is dense. But, since no point of $\mathfrak{A}_0$ cannot belong to $W_\varrho(f)$, so that $\mathfrak{A}_0 \subset \mathscr{C}W_\varrho(f)$ and hence $\mathcal{M}(\theta) \cap \cap \mathfrak{A}_0 = \theta$. This is a contradiction.

2°. Similarly $I(f) \cap \mathscr{C}W_\varrho(f)$ is of category I. Put $\mathcal{M}(\theta) = I(f) \cap \cap \mathscr{C}W_\varrho(f)$ and apply Lemma 2 under the hypothesis that $\mathcal{M}(\theta)$ is of category II. This implies the existence of a subset $\mathcal{M}_0(\theta) \subset \mathcal{M}(\theta)$ of category II such that $\bigcap_{\mathcal{M}_0} \mathscr{C}C_\varrho(f, e^{i\theta}) \neq \theta$. Applying Lemma 1, we see that there is a Fatou arc $\mathfrak{A}_0$ on which $\mathcal{M}_0(\theta)$ is dense. But, evidently no point of $\mathfrak{A}_0$ belongs to $I(f)$, so that $\mathfrak{A}_0 \subset \mathscr{C}I(f)$. We have a contradiction, since $\mathcal{M}(\theta) \cap \mathfrak{A}_0 = \theta$.

**4.** Collingwood [5, 9, 10] has introduced an important notion of a set of maximum indetermination of a function at a frontier point or a set and has obtained systematically many significant theorems on cluster sets.

Let $w = f(z)$ be meromorphic in the unit circle $D$: $|z| < 1$. Let $S$ be a sub-set of $D$ such that $e^{i\theta} \in \bar{S}$. If the cluster set $C_S(f, e^{i\theta})$, on $S$, of $f(z)$ at $e^{i\theta}$ is identical with $C_D(f, e^{i\theta})$, then $S$ is said to be a *set of maximum indetermination* of $f(z)$ at $e^{i\theta}$; and the set $C_S(f, e^{i\theta})$ in this case is *maximal*. Obviously, we can also adopt the corresponding definition in the large provided that $\bar{S}$ intersects the circumference $\Gamma$: $|z| = 1$ and $C_S(f) = C_D(f)$[1].

Now, let $S_0$ be a set in $|z| < 1$ such that (i) the set $\{|z| : z \in S_0\}$ is dense in an interval $r_0 < |z| < 1$ and (ii) given $\eta > 0$ there is a positive number $\delta$ such that $|\arg z| < \eta$ for all $z \in S_0$ belonging to the annulus $1 - \delta < |z| < 1$. We denote by $S_\theta$ the transform of $S_0$ by the rotation $z' = e^{i\theta} z$ so that $S_0$ has $z = e^{i\theta}$ as its only accumulation point on $\Gamma$.

The following theorem is fundamental.

**Theorem 10.** *If $f(z)$ is meromorphic in $|z| < 1$ and if $\{S_\theta\}$ is the family of rotations of a given set $S_0$ which satisfies conditions* (i) *and* (ii), *then for all $e^{i\theta}$ of a residual set on $\Gamma$: $|z| = 1$, $S_\theta$ is a set of maximum indetermination of $f(z)$ at $e^{i\theta}$, i. e.*

$$C_{S_\theta}(f, e^{i\theta}) = C_D(f, e^{i\theta})$$

(Collingwood [9])[2].

*Proof.* Let $\mathcal{M}(\theta)$ be the set of points $e^{i\theta}$ for which

$$C_{S_\theta}(f, e^{i\theta}) \neq C_D(f, e^{i\theta}) .$$

We prove that the hypothesis that $\mathcal{M}(\theta)$ is of category II on $\Gamma$ leads to a contradiction. For this purpose, we shall make use of the spherical metric, considering the Riemann $w$-sphere. Let $\{\varepsilon_n\}$ be a monotonically decreasing sequence of positive numbers such that $\lim_{n\to\infty} \varepsilon_n = 0$. We denote by $C_{S_\theta}(f, e^{i\theta})_{+\varepsilon_n}$ the (closed) set of points whose spherical distances from $C_{S_\theta}(f, e^{i\theta})$ are at most equal to $\varepsilon_n$. Let $E_n$ be the set of points $e^{i\theta}$ for which $C_D(f, e^{i\theta}) - C_{S_\theta}(f, e^{i\theta})_{+\varepsilon_n}$ is not empty. Obviously $\mathcal{M}(\theta) = \bigcup_n E_n$. Now suppose that $\mathcal{M}(\theta)$ is of category II. Then there exists at least one $N$ such that $E_N$ is also of category II. On the Riemann $w$-sphere, we consider a triangulation consisting of a finite number of triangles $\Delta_1, \Delta_2, \ldots, \Delta_m$, each of which has a (spherical) diameter less than $\frac{1}{4} \cdot \varepsilon_N$. For a given $\mu \leqq m$ we denote by $E_{N,\mu}$ the subset of $E_N$ such that for all $e^{i\theta} \in E_{N,\mu}$

$$\Delta_\mu \cap (C_D(f, e^{i\theta}) - C_{S_\theta}(f, e^{i\theta})_{+\varepsilon_N}) \neq \emptyset . \tag{8}$$

Since $E_N = \bigcup_{\mu \leqq m} E_{N,\mu}$, we can find $\mu = M$, such that $E_{N,M}$ is of category II. We sub-divide this set $E_{N,M}$ in the following way. Let $\alpha$ be an arbirarily fixed point of $\Delta_M$. Note that the closed (spherical) circular disc $c_\alpha$:

[1] $\alpha \in C_S(f)$, if there is a sequence of points $\{z_n\}$ such that $|z_n| \to 1$, $z_n \in S$ and $f(z_n) \to \alpha$. It is proved that for any function meromorphic in $|z| < 1$ there exists an isolated set $S$ of points $\{z_n\}$, $|z_n| < |z_{n+1}|$, $\lim |z_n| = 1$ such that $C_S(f) = C_D(f)$ (Collingwood [5]).

[2] The conjecture of a theorem of this type is initially due to Eggleston (see Collingwood [4]).

$[w, \alpha] \leqq \frac{1}{2} \varepsilon_N$ contains $\Delta_M$ in its interior, where $[w, \alpha]$ denotes the spherical distance between $w$ and $\alpha$, and that $c_\alpha \cap C_{S_\theta}(f, e^{i\theta}) = \emptyset$ for every $e^{i\theta} \in E_{N,M}$. Now, for each natural number $p$, we denote by $E_{N,M,p}$ the subset of $E_{N,M}$ such that $[f(z), \alpha] > \frac{1}{2} \varepsilon_N$ for every $z \in S_\theta$, $1 - 2^{-p} < |z| < 1$. Then $E_{N,M} = \bigcup_p E_{N,M,p}$. Since there exists $P$ such that $E_{N,M,P}$ is of category II on $\Gamma$, we can find an arc $\mathfrak{A}$ of $\Gamma$ on which $E_{N,M,P}$ is dense. As $e^{i\theta}$ traverses $\mathfrak{A}$ the set $S_\theta$ sweeps out a set $G$ of circular arcs.

Let $\mathfrak{A}_0$ be an open arc contained in $\mathfrak{A}$ and let $2\eta$ be its angular distance from the two end-points of $\mathfrak{A}$. By (ii), we can find $\delta > 0$ such that in the annulus $1 - \delta < |z| < 1$ all points $z \in S_\theta$ lie in the angle $|\arg z - \theta| < \eta$. Consequently the end points of the circular arcs of the set $G$ contained in this annulus $1 - \delta < |z| < 1$ all lie outside the sector $\Sigma$ bounded by $\mathfrak{A}_0$ and the radii to its end points. Now select a positive number $\delta' < \min(\delta, 2^{-P}, 1 - r_0)$ and denote by $\Sigma_0$ the intersection of $\Sigma$ with the annulus $1 - \delta' < |z| < 1$. By (i), the set of points $G \cap \Sigma_0$ is dense in $\Sigma_0$ and for every $z \in (G \cap \Sigma_0)$ we have $[f(z), \alpha] > \frac{1}{2} \varepsilon_N$ provided that $z$ belongs to some $S_\theta$, $e^{i\theta} \in E_{N,M,P}$. Accordingly, by continuity, $[f(z), \alpha] \geqq \frac{1}{2} \varepsilon_N$ for all $z \in \Sigma_0$. From this it follows that $\Delta_M \cap C_D(f, e^{i\theta'}) = \emptyset$ for all $e^{i\theta'} \in \mathfrak{A}_0$, since $\alpha \in \Delta_M$ and $\Delta_M$ is of diameter less than $\frac{1}{4} \varepsilon_N$. But, since $E_{N,M,P}$ is dense on $\mathfrak{A}$ and therefore also on $\mathfrak{A}_0$, we can find a point $e^{i\theta'} \in \mathfrak{A}_0 \cap E_{N,M,P}$ so that for this value of $\theta'$

$$\Delta_M \cap (C_D(f, e^{i\theta'}) - C_{S_{\theta'}}(f, e^{i\theta'})_{+\varepsilon_N}) \neq \emptyset$$

and therefore $\Delta_M \cap C_D(f, e^{i\theta'}) \neq \emptyset$. This is a contradiction.

Remark. The only property of analytic functions used in the proof is the property of being a continuous mapping.

Now we consider the chordal cluster set $C_{\varrho(\varphi)}(f, e^{i\theta})$. Obviously $\varrho(\varphi)$ satisfies the conditions on $S_\theta$. As an immediate consequence of Theorem 10, we have

**Corollary 5.** *Let $f(z)$ be meromorphic in $D$: $|z| < 1$ and let $\{\varphi_n\}$ be any enumerable set in the open interval $(-\pi/2, \pi/2)$. Then the set of points $e^{i\theta}$ for which*

$$C_D(f, e^{i\theta}) = C_{\varrho(\varphi_1)}(f, e^{i\theta}) = \cdots = C_{\varrho(\varphi_n)}(f, e^{i\theta}) = \cdots$$

*is residual on $\Gamma$: $|z| = 1$* (Collingwood [4, 9]).

For the set of points $e^{i\theta}$ in which any one of the sets $C_{\varrho(\varphi_n)}(f, e^{i\theta})$ is not maximal is of category I on $\Gamma$.

**5.** We shall explain on certain classes of singularities defined by cluster sets and their mutual relations.

Let $w = f(z)$ be meromorphic in $D$: $|z| < 1$. We have already introduced $F(f)$ = the set of all Fatou points on $\Gamma$: $|z| = 1$, $I(f)$ = the set of all Plessner points and $W_\varrho(f)$ = the set of points $e^{i\theta}$ for which the radial cluster set $C_\varrho(f, e^{i\theta})$ is total. We now introduce some new notations.

$J(f)$ = the set of points $e^{i\theta} \in \Gamma$ such that for every Stolz angle $\Delta$ at $e^{i\theta}$ the equation $C_\Delta(f, e^{i\theta}) = C_D(f, e^{i\theta})$ is satisfied.

$K(f)$ = the set of points $e^{i\theta} \in \Gamma$ such that given any two Stolz angles $\Delta_1$, $\Delta_2$ at $e^{i\theta}$, whether overlapping or not, the equation $C_{\Delta_1}(f, e^{i\theta}) = C_{\Delta_2}(f, e^{i\theta})$ is satisfied.

$W(f)$ = the set of points $e^{i\theta}$ for which $C_D(f, e^{i\theta})$ is total.

$W_{\varrho(\varphi)}(f)$ = the set of points $e^{i\theta}$ for which the chordal cluster set $C_{\varrho(\varphi)}(f, e^{i\theta})$ is total.

$D(f)$ = the set of points $e^{i\theta}$ for which $C_D(f, e^{i\theta})$ is degenerate.

$D_{\varrho(\varphi)}(f)$ = the set of points $e^{i\theta}$ for which $C_{\varrho(\varphi)}(f, e^{i\theta})$ is degenerate.

More generally, if $S_\theta$ satisfies the conditions of Theorem 10, $W_{S\theta}(f, e^{i\theta})$ denotes the set $\{e^{i\theta}\}$ for which $C_{S\theta}(f, e^{i\theta})$ is total; similarly $D_{S\theta}(f, e^{i\theta})$ denotes the set for which $C_{S\theta}(f, e^{i\theta})$ is degenerate. It is obvious that $J(f) \subset K(f)$ and $F(f) \cup I(f) \subset K(f)$.

**Theorem 11.** *If $f(z)$ is meromorphic in $|z| < 1$, then the sets $J(f)$ and $K(f)$ are residual on $\Gamma$* (COLLINGWOOD [9]).

*Proof.* Take the set $\{\varphi_n\}$ of Corollary 5 to be a dense countable set in the interval $(-\pi/2, \pi/2)$ and denote by $\mathcal{M}$ the residual set of points $e^{i\theta}$ for which $C_{\varrho(\varphi_n)}(f, e^{i\theta}) = C_D(f, e^{i\theta})$ is satisfied for every $n$. Since every Stolz angle $\Delta$ at $e^{i\theta} \in \mathcal{M}$ contains a chord $\varrho(\varphi_n)$, we have $C_{\varrho(\varphi_n)}(f, e^{i\theta}) \subset \subset C_\Delta(f, e^{i\theta})$ and therefore $C_\Delta(f, e^{i\theta}) = C_D(f, e^{i\theta})$. Consequently $\mathcal{M} \subset J(f)$ and so $J(f)$ is residual on $\Gamma$. Since $J(f) \subset K(f)$, it follows that $K(f)$ is also residual on $\Gamma$.

**Theorem 12.** *For any function $f(z)$ meromorphic in $|z| < 1$, the set $K(f)$ is residual and of measure $2\pi$ on $\Gamma$.*

*Proof.* By Plessner's theorem $F(f) \cup I(f)$ is of measure $2\pi$ and $F(f) \cup I(f) \subset K(f)$.

Remark. It is not true in general that $F(f) \cup I(f)$ is residual on $\Gamma$. Consider a Blaschke product $f(z)$ whose zero-points are so distributed that the closure of their union contains $\Gamma$. In this case, $I(f)$ is empty and $C_D(f, e^{i\theta})$ is the closed disc $|w| \leqq 1$ for all $e^{i\theta}$ of $\Gamma$; therefore $F(f) \subset \mathscr{C} J(f)$ which is of category I by Theorem 11.

**Theorem 13.** *Let $f(z)$ be meromorphic in $|z| < 1$. Then any two of the sets $W(f)$, $W_{S\theta}(f)$ and $I(f)$ where $S_\theta$ satisfies the conditions of* Theorem 10, *differ at most by a set of category I on $\Gamma$* (COLLINGWOOD [9]).[1]

*Proof.* By Theorem 10, $W(f) \cap \mathscr{C} W_{S\theta}(f)$ is of category I and since $W_{S\theta}(f) \subset W(f)$ it follows that $W(f)$ differs from $W_{S_\theta}(f)$ by a set of cate-

[1] Theorem 13 contains Theorem 8.

gory I. On the other hand $W(f) \cap \mathscr{C} I(f) \subset \mathscr{C} J(f)$ and $I(f) \subset W(f)$. Since $\mathscr{C} J(f)$ is of category I, it follows that $W(f)$ differs from $I(f)$ by a set of category I.

**Theorem 14.** *Let $f(z)$ be meromorphic in $|z| < 1$. Then any two of the sets $D(f)$, $D_{S_\theta}(f)$ and $F(f)$, where $S_\theta$ satisfies the conditions of* Theorem 10, *differ at most by a set of category* I *on $\Gamma$* (COLLINGWOOD [9]).[1]

*Proof.* By Theorem 10, $D_{S_\theta}(f) \cap \mathscr{C} D(f)$ is of category I and since $D(f) \subset D_{S_\theta}(f)$ it follows that $D(f)$ differs from $D_{S_\theta}(f)$ only by a set of category I. The theorem is proved, since $D(f) \subset F(f) \subset D_\varrho(f)$.

**6.** We state some properties of $K(f)$ and $J(f)$ without proofs. Let $w = f(z)$ be meromorphic in $D$: $|z| < 1$. We define the *outer angular cluster set* $C_{\mathscr{A}}(f, e^{i\theta})$ of $f(z)$ at $e^{i\theta}$ as the union

$$C_{\mathscr{A}}(f, e^{i\theta}) = \bigcup_{\Delta} C_\Delta(f, e^{i\theta})$$

taken over all Stolz angles $\Delta$ at $e^{i\theta}$.

By using the idea of the proof of Theorem 10, we can prove

**Theorem 15.** *If $f(z)$ is meromorphic in $|z| < 1$ and if $e^{i\theta} \in K(f)$, then $C_{\varrho(\varphi)}(f, e^{i\theta}) = C_{\mathscr{A}}(f, e^{i\theta})$ for a set of values $\varphi$ residual in the open interval* $(-\pi/2, \pi/2)$ (COLLINGWOOD [9]).

**Corollary 6.** *If $f(z)$ is meromorphic in $|z| < 1$ and if $e^{i\theta} \in J(f)$, then $C_{\varrho(\varphi)}(f, e^{i\theta}) = C_D(f, e^{i\theta})$ for a set of values of $\varphi$ residual in* $(-\pi/2, \pi/2)$.

For, in $J(f)$ we have $C_{\mathscr{A}}(f, e^{i\theta}) = C_D(f, e^{i\theta})$ and $J(f) \subset K(f)$.

Denote now by $\Pi(f, e^{i\theta})$ the intersection of the cluster sets $C_\lambda(f, e^{i\theta})$ on all possible curves $\lambda$ in $|z| < 1$ ending at $e^{i\theta}$.[2] Then we have

**Theorem 16.** *If $f(z)$ is meromorphic in $|z| < 1$ and if $C(f) = \mathscr{F} C(f)$* [3], *then*

$$C_D(f, e^{i\theta}) = \Pi(f, e^{i\theta})$$

*for all $e^{i\theta} \in J(f)$* (COLLINGWOOD [9])[4].

**7.** We now suppose that $w = f(z)$ is an arbitrary complex-valued function defined in $D$: $|z| < 1$. Denote by $M_{\eta r}$ the closure of the union $\bigcup_{\theta'} C_D(f, e^{i\theta'})$ for all $\theta'$ in the interval $(\theta - \eta, \theta)$ and write

$$C_{\Gamma r}(f, e^{i\theta}) = \bigcap_{\eta > 0} M_{\eta r}$$

which is called the *right-hand boundary cluster set* of $f(z)$ at $z = e^{i\theta}$. Similarly we can define the *left-hand boundary cluster set* $C_{\Gamma l}(f, e^{i\theta})$ of

---

[1] In the special case where $S_\theta$ is the radius $\varrho$ to $e^{i\theta}$, this theorem has been proved by EGGLESTON [2] and COLLINGWOOD [7].

[2] We call $\Pi(f, e^{i\theta})$ the principal subset of $C(f, e^{i\theta})$ and a value $\alpha \in \Pi(f, e^{i\theta})$ a principal value at $e^{i\theta}$.

[3] $\mathscr{F} C(f)$ denotes the frontier of $C(f)$.

[4] In the special case of univalent functions this theorem gives, as a corollary, the complete solution of an old problem of CARATHÉODORY on prime ends (cf. COLLINGWOOD [8]).

$f(z)$ at $z = e^{i\theta}$. The complete boundary cluster set $C_\Gamma(f, e^{i\theta})$ satisfies

$$C_\Gamma(f, e^{i\theta}) = C_{\Gamma_r}(f, e^{i\theta}) \cup C_{\Gamma_l}(f, e^{i\theta}) .$$

By an elaboration of the proof of Theorem 10, COLLINGWOOD has obtained the following beautiful result.

**Theorem 17.** *Suppose that $f(z)$ is an arbitrary complex-valued function in $D$: $|z| < 1$. Then the boundary cluster sets $C_{\Gamma_r}(f, e^{i\theta})$, $C_{\Gamma_l}(f, e^{i\theta})$ and $C_\Gamma(f, e^{i\theta})$ are all maximal in a residual set on $\Gamma$: $|z| = 1$; i. e. $C_{\Gamma_r}(f, e^{i\theta}) = C_D(f, e^{i\theta})$ and similarly for $C_{\Gamma_l}(f, e^{i\theta})$ and $C_\Gamma(f, e^{i\theta})$* (COLLINGWOOD [9], **[10]**)[1].

Remark. This theorem is closely related to Bagemihl's result on ambiguous points of an arbitrary function (see Theorem 8, § 1, II). From this theorem and Bagemihl's result, it follows that for any arbitrary function $w = f(z)$ in $D$: $|z| < 1$ the intersection $C_{\Gamma_r}(f, e^{i\theta}) \cap C_{\Gamma_l}(f, e^{i\theta})$ is not empty except perhaps at a countable set of points $e^{i\theta}$ of $\Gamma$ and maximal except perhaps in a set of category I on $\Gamma$.

## § 4. Boundary behaviour of meromorphic functions

1. We have already stated, in § 3, the following

**Theorem 1.** *Let $w = f(z)$ be meromorphic in the unit cirle $D$: $|z| < 1$. Then, almost every point $e^{i\theta}$ of $\Gamma$: $|z| = 1$ is either a Fatou point or a Plessner point* (PLESSNER [1]).

For the sake of completeness, we state now an outline of the proof. Let $\Delta_\alpha(\theta)$ be a Stolz angle with magnitude $\alpha$ $(0 < \alpha < \pi)$ and vertex $z = 1$. We denote by $\Delta_\alpha(\theta)$ the transform of $\Delta_\alpha(0)$ under the rotation $z' = e^{i\theta} z$ $(0 < \theta < 2\pi)$.

**Lemma 1.** *Let $\mathcal{M}$ be a measurable set of positive measure on $\Gamma$: $|z| = 1$. Suppose that the cluster set $C_{\Delta_\alpha(\theta)}(f, e^{i\theta})$ is sub-total for every point $e^{i\theta} \in \mathcal{M}$. Then, $\mathcal{M}$ includes a set $\mathcal{M}_1$ of positive measure which consists of Fatou points.*

*Proof.* We construct on the Riemann $w$-sphere a sequence of finite triangular lattices $l_1, l_2, \ldots, l_n, \ldots$, each lattice being a sub-division of its predecessor, such that the maximum of diameters of triangles in the lattice $l_n$ is less than $1/2^n$. We denote the individual triangles in the lattice $l_n$ by $\gamma_{n,1}, \gamma_{n,2}, \ldots, \gamma_{n,m(n)}$. Denote by $\Gamma_n(\theta)$ the set of points $e^{i\theta} \in \mathcal{M}$ such that $n$ is the smallest number for which $\mathscr{C} C_{\Delta_\alpha(\theta)}(f, e^{i\theta})$ contains at least one of the triangles $\gamma_{n,\nu} (1 \leqq \nu \leqq m(n))$. We now sub-divide the set $\Gamma_n(\theta)$, assigning to each triangle $\gamma_{n,1}, \gamma_{n,2}, \ldots, \gamma_{n,m(n)}$ all those values of $e^{i\theta} \in \Gamma_n(\theta)$ for which $\gamma_{n,\nu}$, $1 \leqq \nu \leqq m(n)$, is contained in $\mathscr{C} C_{\Delta_\alpha(\theta)}$, and we denote the corresponding subsets of $\Gamma_n(\theta)$ by $E_{n,\nu}(\theta)$, $1 \leqq \nu \leqq m(n)$. Obviously $\mathcal{M} = \bigcup_n \bigcup_{1 \leqq \nu \leqq m(n)} E_{n,\nu}(\theta)$. Since $E_{n,\nu}(\theta)$ is measurable and $\mathcal{M}$ is of positive measure, there exists at least one $E_{n,\nu}(\theta)$, $n < \infty$, $\nu \leqq m(n)$, say,

[1] See the reconstructed proof in COLLINGWOOD [10].

$E_{j,k}$ which has positive measure.[1] By using a suitable linear transformation $W = \varphi(w)$, we map the complement of $\gamma_{j,k}$ onto a domain in the disc $|W| < M$. Consider the composed function $W = f^*(z) \equiv \varphi[f(z)]$. Then, $f^*(z)$ is meromorphic in the unit circle and possesses the property that for every point $e^{i\theta} \in E_{j,k}$ the cluster set $C_{\Delta_\alpha(\theta)}(f^*, e^{i\theta})$ lies in the interior of $|W| < M$. Let $\{r_n\}$ be a monotonically increasing sequence of positive numbers converging to unity. For every $n$, let $e^{i\theta} \in A_n$, if $|f^*(z)| < M$ in $\Delta_\alpha(\theta)$ for $|z| \geqq r_n$. Clearly, $A_1 \subset A_2 \subset \ldots \subset A_n \subset \ldots$ and $E_{j,k} = \bigcup\limits_n A_n$; hence $mE_{j,k} = \lim\limits_{n\to\infty} mA_n$. Consequently, there exists a set $A_{n_1}$ with $mA_{n_1} > 0$ and hence we can select a perfect subset $\mathcal{M}_1$ of $A_{n_1}$ of positive measure. Put $\varrho = r_{n_1}$. Now, we notice that the perfect set $\mathcal{M}_1$ has the property that for every $e^{i\theta} \in \mathcal{M}_1$, we have $|f^*(z)| < M$ in the intersection $D_{\alpha,\varrho}(\theta)$ of $\Delta_\alpha(\theta)$ and $\{z: \varrho \leqq |z| < 1\}$. Denote by $G$ the domain $\{\bigcup\limits_{\mathcal{M}_1} D_{\alpha,\varrho}(\theta)\} \cup \{z: |z| < \varrho\}$. By a geometric consideration, we see that $G$ is a simply connected domain bounded by a rectifiable Jordan curve $K$. Let $a_1, a_2, \ldots, a_p$ be all the poles of $f^*(z)$ in $|z| \leqq \varrho$ and form the function

$$g(z) = (z - a_1)(z - a_2) \ldots (z - a_p) f^*(z)$$

which is regular and bounded in $G$. Then, by an extension of Fatou's theorem, $g(z)$ has angular limits for almost all points of $K$. Obviously this holds also for $f(z)$. Notice that the unit circumference $\Gamma$: $|z| = 1$ and the rectifiable Jordan curve $K$ have the set $\mathcal{M}_1$ in common. For almost every point $e^{i\theta} \in \mathcal{M}_1$, $K$ has a tangent which concides with the tangent of $\Gamma$ at $e^{i\theta}$, since $\mathcal{M}_1$ is perfect. Consequently, the angular limits of $f(z)$ in $D$: $|z| < 1$ and in $G$ are identical for almost every point $e^{i\theta} \in \mathcal{M}_1$. Thus, it is proved that almost every point of $\mathcal{M}_1$ is a Fatou point.

Now, we state the proof of Plessner's theorem. We divide the unit circumference $\Gamma$: $|z| = 1$ into three parts:

$$\Gamma = F(f) \cup \Gamma_2^{(\alpha)} \cup \Gamma_3^{(\alpha)},$$

where $F(f)$ is the set of all Fatou points, $\Gamma_2^{(\alpha)}$ is the set of all points $e^{i\theta}$ for which the cluster set $C_{\Delta_\alpha(\theta)}$ is total and $\Gamma_3^{(\alpha)}$ is the complement of $F(f) \cup \Gamma_2^{(\alpha)}$ with respect to $\Gamma$. By the lemma, $m\Gamma_3^{(\alpha)} = 0$. Otherwise, $\Gamma_3^{(\alpha)}$ would contain a set of positive measure consisting of Fatou points; this is a contradiction. Now, consider a sequence of Stolz angles $\{\Delta_{\alpha_n}(\theta)\}$ with vertex $e^{i\theta}$ and with the property that for any Stolz angle $\Delta(\theta)$ there exists an angle $\Delta_{\alpha_n}(\theta)$ which is contained in $\Delta(\theta)$. For every $\Delta_{\alpha_n}(\theta)$, we have the corresponding decomposition

$$\Gamma = F(f) \cup \Gamma_2^{(\alpha_n)} \cup \Gamma_3^{(\alpha_n)}, \quad m\Gamma_3^{(\alpha_n)} = 0.$$

[1] This method has been used by COLLINGWOOD to prove Lemma 2, § 3.

Put $\Gamma_3 = \bigcup_n \Gamma_3^{(\alpha_n)}$ and $\Gamma_2 = \bigcap_n \Gamma_2^{(\alpha_n)}$. Then $\Gamma = F(f) \cup \Gamma_2 \cup \Gamma_3$. It is obvious that $m\Gamma_3 = 0$ and $\Gamma_2$ consists of Plessner points; i. e., $\Gamma_2 = I(f)$.

2. We prove

**Theorem 2.** (LUSIN and PRIVALOFF [1] and PLESSNER [1]). *Let $w = f(z)$ be meromorphic in the unit circle $D$: $|z| < 1$. Let $\mathcal{M}$ be a set of positive measure on $\Gamma$: $|z| = 1$. Suppose that $f(z)$ has angular limit $\gamma$ at every point $e^{i\theta} \in \mathcal{M}$, where $\gamma$ is a fixed complex number. Then, $f(z)$ reduces to a constant*[1].

*Proof.* Obviously $f(z)$ satisfies the condition of the preceding lemma. The function $f^*(z)$ in the proof of the lemma has a fixed angular limit $\beta (= \varphi(\gamma))$ for every point $e^{i\theta}$ of the perfect set $\mathcal{M}_1$. Consider now the function

$$g(z) = (f^*(z) - \beta)(z - a_1)(z - a_2) \ldots (z - a_p)$$

in the domain $G$ bounded by a rectifiable Jordan curve $K$ (see the proof of Lemma 1). Then $g(z)$ is regular and bounded in $G$, and has angular limit 0 at almost all points of $\mathcal{M}_1$ on $K$. Applying a theorem of F. and M. RIESZ, $g(z)$ is identically zero and therefore $f(z)$ reduces to a constant.

3. K. E. MEIER [2] has sharpened Theorem 1. Let $w = f(z)$ be meromorphic in the unit circle $D$: $|z| < 1$. If there are two chords $\Lambda_1$, $\Lambda_2$ terminating at $e^{i\theta}$ such that

$$\gamma \notin \{C_{\Lambda_1}(f, e^{i\theta}) \cup C_{\Lambda_2}(f, e^{i\theta})\},$$

then we say that the value $\gamma$ belongs to $R^*(f, e^{i\theta})$. Evidently, $R^*(f, e^{i\theta})$ is an open set which may be empty[2].

**Theorem 3.** *Let $w = f(z)$ be meromorphic in the unit circle $D$: $|z| < 1$. Let $\mathcal{M}$ be a measurable set of positive measure on $\Gamma$: $|z| = 1$ such that for every $e^{i\theta} \in \mathcal{M}$, $R^*(f, e^{i\theta})$ contains a fixed point $\gamma$. Then, for almost every point $e^{i\theta}$ of $\mathcal{M}$, either $f(z)$ has an angular limit or $f(z)$ assumes the value $\gamma$ infinitely often in any Stolz angle $\Delta(\theta)$ with vertex $e^{i\theta}$* (MEIER [2]).

To prove this theorem, we make use of a lemma analogous to Lemma 1.

**Lemma 2.** *Let $\mathcal{M}$ be a measurable set of positive measure on $\Gamma$: $|z| = 1$. Suppose that $f(z)$ assumes a fixed value $\gamma$, belonging to $R^*(f, e^{i\theta})$, finitely often in the Stolz angle $\Delta_\alpha(\theta)$, defined in* Lemma 1, *for every $e^{i\theta} \in \mathcal{M}$. Then $A = \mathcal{M} \cap I(f)$ is of measure zero, where $I(f)$ denotes the set of Plessner points.*

[1] TSUJI [7] has extended this theorem as follows: Let $w = f(z)$ be meromorphic in $D$: $|z| < 1$. Let $\mathcal{M}$ be a set of positive measure on $\Gamma$: $|z| = 1$ and $E_w$ be a closed set of capacity zero in the $w$-plane. Suppose that the angular cluster set $C_\Delta(f, e^{i\theta})$ of $f(z)$ at every point $e^{i\theta} \in \mathcal{M}$ is contained in $E_w$. Then, $f(z)$ reduces to a constant.

[2] Let $\psi(w)$ be a meromorphic function for which the cluster set at $w = \infty$ on every simple curve which tends to infinity is total. Consider the function

$$f(z) = \psi\left(\frac{1+z}{1-z}\right).$$

Then $R^*(f, 1)$ is empty. Cf. BAGEMIHL-SEIDEL [1], p. 1072.

*Proof.* Suppose that $A = \mathscr{M} \cap I(f)$ is of positive measure. Let $\{r_n\}$ be a monotonically increasing sequence of positive numbers converging to unity. Let $e^{i\theta} \in A_n$, if $f(z) \neq \gamma$ in $\Delta_\alpha(\theta)$ with vertex $e^{i\theta} \in A$ for $|z| > r_n$. Then $A_1 \subset A_2 \subset \ldots \subset A_n \subset \ldots$ and $A = \bigcup_n A_n$; hence $\lim_{n\to\infty} mA_n = mA$. Consequently, there exists a set $A_{n_1}$ of positive measure. Put $r_{n_1} = \varrho$. Let $\mathscr{M}_1$ be a perfect subset of $A_{n_1}$ such that $m\mathscr{M}_1 > 0$. We construct the domain $G$: $\{\bigcup_{\mathscr{M}_1} D_{\alpha,\varrho}(\theta)\} \cup \{z : |z| < \varrho\}$ in Lemma 1. Then, $G$ is a simply connected domain bounded by a rectifiable Jordan curve $K$. Notice that for almost every point $e^{i\theta} \in \mathscr{M}_1$, $K$ has a tangent which coincides with the tangent of $\Gamma$: $|z| = 1$ at $e^{i\theta}$. Now, let $e^{i\theta}$ be such a point of $\mathscr{M}_1$. If $\eta (> 0)$ is sufficiently small, the domain $\Delta_{12}$ bounded by two chords $\Lambda_1$, $\Lambda_2$, terminating at $e^{i\theta}$, and $|z - e^{i\theta}| = \eta$ is completely contained in $G$. Since $e^{i\theta}$ is a Plessner point, $C_{\Delta_{12}}(f, e^{i\theta})$ is total. On the other hand, $f(z) \neq \gamma$ in $G$ for $|z| > \varrho$ and $\gamma \notin \{C_{\Lambda_1}(f, e^{i\theta}) \cup C_{\Lambda_2}(f, e^{i\theta})\}$. By a well-known theorem[1] on cluster sets, there exists an asymptotic path $L$ inside $\Delta_{12}$ terminating at $e^{i\theta}$ along which $f(z)$ converges to $\gamma$. Consequently $C_{\Lambda_1}(f, e^{i\theta}) \cap C_L(f, e^{i\theta}) = \emptyset$; that is, $e^{i\theta}$ is an ambiguous point of $f(z)$ in the sense of BAGEMIHL and SEIDEL. By Bagemihl's theorem[2], $f(z)$ has at most a countable number of ambiguous points. This is a contradiction, since $\mathscr{M}_1$ is of positive measure.

Now, we prove Theorem 3. We divide $\mathscr{M}$ into three parts:

$$\mathscr{M} = \{F(f) \cap \mathscr{M}\} \cup \mathscr{M}_2^{(\alpha)} \cup \mathscr{M}_3^{(\alpha)}$$

where $\mathscr{M}_2^{(\alpha)}$ denotes the set of all points $e^{i\theta}$ of $\mathscr{M}$ for which $f(z)$ assumes the value $\gamma \in R^*(f, e^{i\theta})$ infinitely often in $\Delta_\alpha(\theta)$ and $\mathscr{M}_3^{(\alpha)}$ the complement of $\{F(f) \cap \mathscr{M}\} \cup \mathscr{M}_2^{(\alpha)}$ with respect to $\mathscr{M}$. By Lemma 2, $\mathscr{M}_3^{(\alpha)}$ is of measure zero. Consider a sequence of Stolz angles $\{\Delta_{\alpha_n}(\theta)\}$ with vertex $e^{i\theta}$ such that for any $\Delta(\theta)$ there exists an angle $\Delta_{\alpha_n}(\theta)$ which is contained in $\Delta(\theta)$. For every $\Delta_{\alpha_n}(\theta)$, we have the corresponding decomposition

$$\mathscr{M} = \{F(f) \cap \mathscr{M}\} \cup \mathscr{M}_2^{(\alpha_n)} \cup \mathscr{M}_3^{(\alpha_n)}, \quad m\mathscr{M}_3^{(\alpha_n)} = 0\,.$$

Put $\mathscr{M}_3 = \bigcup_n \mathscr{M}_3^{(\alpha_n)}$ and $\mathscr{M}_2 = \bigcap_n \mathscr{M}_2^{(\alpha_n)}$. Then $\mathscr{M} = \{\mathscr{M} \cap F(f)\} \cup \mathscr{M}_2 \cup \mathscr{M}_3$. Obviously $m\mathscr{M}_3 = 0$ and $\mathscr{M}_2$ is the set of points $e^{i\theta}$ for which $f(z)$ assumes $\gamma \in R^*(f, e^{i\theta})$ infinitely often in any Stolz angle $\Delta(\theta)$ with vertex $e^{i\theta}$. Thus the conclusion of the theorem is proved.

**4. Some applications.** Let $\mathscr{M}$ be a measurable set of positive measure on $\Gamma$: $|z| = 1$.

As an immediate consequence of Theorem 3, we have

**Theorem 4.** *Let $w = f(z)$ be meromorphic in $|z| < 1$. Suppose that for every point $e^{i\theta} \in \mathscr{M}$, there exist two chords $\Lambda_1$, $\Lambda_2$, terminating at $e^{i\theta}$, on which*

[1] This means the special case where $E$ consists of a single point, of Theorem 1, § 3, II.

[2] Cf. Theorem 8, § 1.

*$f(z)$ is bounded. Then, for almost every point $e^{i\theta} \in \mathcal{M}$, $f(z)$ has a finite angular limit or $f(z)$ assumes $\infty$ infinitely often in any Stolz angle $\Delta(\theta)$ with vertex at $e^{i\theta}$* (MEIER [2]).

**Corollary 1.** *Let $f(z)$ be regular in the unit circle $D: |z| < 1$. Suppose that for every point $e^{i\theta} \in \mathcal{M}$, there are two chords $\Lambda_1$, $\Lambda_2$ terminating at $e^{i\theta}$ on which $f(z)$ is bounded. Then, $f(z)$ has a finite angular limit at almost every point $e^{i\theta}$ of $\mathcal{M}$* (MEIER [1]).

Remark. In this corollary, it is possible to replace "two chords" by "two arbitrary simple arcs" provided that there exists a Stolz angle $\Delta(\theta)$ with vertex $e^{i\theta}$ between two arcs $\Lambda_1$ and $\Lambda_2$. Because, if $e^{i\theta}$ is a Plessner point, then $C_{\Delta(\theta)}(f, e^{i\theta})$ is total and hence there exists an asymptotic path $L$, terminating at $e^{i\theta}$, along which $f(z)$ converges to $\infty$. Consequently, the point $e^{i\theta}$ is an ambiguous point of $f(z)$. We have only to note that there are at most countably many ambiguous points of $f(z)$ by Bagemihl's theorem.

Applying Theorem 2 and Theorem 3, we obtain

**Theorem 5.** *Let $f(z)$ be meromorphic in the unit circle $D: |z| < 1$. Suppose that for every $e^{i\theta} \in \mathcal{M}$, there are two chords $\Lambda_1$, $\Lambda_2$ terminating at $e^{i\theta}$ along which $f(z)$ converges to $0$ as $z \to e^{i\theta}$. Then, either $f(z) \equiv 0$ or $f(z)$ takes every value $\gamma$ different from $0$ infinitely often in any Stolz angle $\Delta(\theta)$ for almost every $e^{i\theta} \in \mathcal{M}$* (MEIER [2]).

**5.** In connection with Meier's results, the following is important.

**Theorem 6.** *There exists a function $f(z)$ regular in $D: |z| < 1$ such that for every $\theta$ in the interval $0 \leqq \theta \leqq 2\pi$, $f(z) \to \infty$ as $z \to e^{i\theta}$ along all chords $\varrho(\varphi)$ of $|z| < 1$ ending at $e^{i\theta}$ except perhaps a set of values $\varphi$ of measure zero in the open interval $(-\pi/2, \pi/2)$* (BAGEMIHL-SEIDEL [8]).

*Proof.* Let $n_j = 3^j$ $(j = 1, 2, \ldots)$ and define $f(z)$ to be the infinite product

$$f(z) = \prod_{j=1}^{\infty} \left\{ 1 - \left( \frac{z}{1 - n_j^{-1}} \right)^{n_j^2} \right\} \tag{1}$$

for $|z| < 1$. This product evidently converges absolutely and uniformly in any compact subset of $D$ and therefore $f(z)$ is regular in $D$. The zeros of $f(z)$ are

$$z_{j,\nu} = \left( 1 - \frac{1}{n_j} \right) e^{2\pi i \nu / n_j^2} \quad (j = 1, 2, \ldots; \quad \nu = 0, 1, \ldots, n_j^2 - 1). \tag{2}$$

Let $\varrho_j = 1/n_j^4$ $(j = 1, 2, \ldots)$ and

$$K_{j,\nu} = \{z : |z - z_{j,\nu}| < \varrho_j\} \quad (j = 1, 2, \ldots; \quad \nu = 0, 1, \ldots, n_j^2 - 1). \tag{3}$$

Denote by $G$ the set obtained by excluding all the discs $K_{j,\nu}$ from $|z| < 1$. Then, we can show that $f(z) \to \infty$ as $|z| \to 1$ provided that $z$ lies in $G$[1].

[1] See the detail of the proof in BAGEMIHL-SEIDEL [8]. For related theorems, cf. BAGEMIHL-ERDÖS-SEIDEL [1].

Now, let $e^{i\theta}$ be any point of $\Gamma$: $|z| = 1$ and fix $k$. The angle subtended at $e^{i\theta}$ by any one of the discs $K_{k,\nu}$ $(\nu = 0, 1, \ldots, n_k^2 - 1)$ is at most

$$2 \arcsin (n_k \varrho_k) = 2 \arcsin (1/n_k^3) < c/n_k^3 ,$$

where $c$ is a positive universal constant. Since there are $n_k^2$ of these discs, the sum of the angles subtended at $e^{i\theta}$ by all these discs is less than $c/n_k$. Since $\sum_{k=1}^{\infty} n_k^{-1}$ is convergent, it follows that almost every chord $\varrho(\varphi)$ ending at $e^{i\theta}$ is such that all its points lying in a sufficiently small neighborhood of $e^{i\theta}$ belong to $G$ and this completes the proof[1].

Setting $g(z) = 1/f(z)$, we have

**Corollary 2**[2]. *There exists a non-constant function $g(z)$ meromorphic in $D$: $|z| < 1$ such that, for every $\theta$ in the interval $0 \leqq \theta \leqq 2\pi$, $g(z) \to 0$ as $z \to e^{i\theta}$ along almost all chords $\varrho(\varphi)$ of $|z| < 1$ terminating at $e^{i\theta}$.*

Remark. From Theorem 4 (Meier) and Theorem 2 (Lusin-Privaloff) it follows that Corollary 2 does not remain valid if we replace the word "meromorphic" therein by the word "regular". More precisely, the function $g(z)$ in Corollary 2 has the following properties:

(i) The set $F(g)$ of Fatou points is of measure zero. (ii) The set $I(g)$ of Plessner points is of measure $2\pi$. (iii) If $e^{i\theta} \in I(g)$ and if $e^{i\theta}$ is not an ambiguous point, then $g(z)$ takes every value $\alpha \neq 0$ infinitely often in any Stolz angle $\Delta(\theta)$ with vertex $e^{i\theta}$[3]. (iv) For every $e^{i\theta}$ of $\Gamma$: $|z| = 1$, $C_D(g, e^{i\theta})$ is total. (v) For every $e^{i\theta}$ of $\Gamma$, $R_D(g, e^{i\theta})$ is the complement of $\{0\}$ with respect to the whole $w$-plane. Now, for every point $e^{i\theta}$ of $\Gamma$, we choose a chord $\lambda_\theta$, ending at $e^{i\theta}$, such that $C_{\lambda_\theta}(g, e^{i\theta}) = \{0\}$, then we have $C_D(g, e^{i\theta}) \neq C_{\lambda_\theta}(g, e^{i\theta})$ for all points $e^{i\theta}$ of $\Gamma$. This implies that Theorem 10 (Collingwood), § 3, does not remain valid if we replace $\{S_\theta\}$ therein by an arbitrary family $\{\lambda_\theta\}$.

Using the simple structure of the domain $G$ in the proof of Theorem 6, we can prove

**Theorem 7.** *Let $f(z)$ be the regular function in $D$: $|z| < 1$ defined by* (1). *Then there exists a set of spirals $\sigma_t$ $(0 \leqq t < 1)$ in $|z| < 1$, each of which approaches $\Gamma$: $|z| = 1$ asymptotically, such that every point of $D$ belongs to*

[1] From the structure of $G$, it is easy to see that there exists a sequence of circumferences $|z| = r_k$, $0 < r_k < 1$, $\lim_{k\to\infty} r_k = 1$ such that, putting $m_k = \min_{|z| = r_k} |f(z)|$, we have $\lim_{k\to\infty} m_k = \infty$ (cf. also Bagemihl-Erdös-Seidel [1], p. 139).

[2] This result was announced in Bagemihl-Seidel [1], pp. 1073—1074.

[3] Suppose that $\alpha \neq 0$ is taken by $g(z)$ only finitely often in $\Delta(\theta)$. Obviously $\Delta(\theta)$ contains a Stolz angle $\Delta$ at $e^{i\theta}$ such that $g(z) \to 0$ as $z \to e^{i\theta}$ along the two sides of $\Delta$. Since $C_\Delta(f, e^{i\theta})$ is total, $\alpha$ must be an asymptotic value on a path inside $\Delta$ and therefore $e^{i\theta}$ will be an ambiguous point of $g(z)$. Contradiction.

*one, and only one, of these spirals, and, for every t,*

$$\lim_{\substack{|z|\to 1\\ z\in\sigma_t}} f(z) = \infty \tag{4}$$

(BAGEMIHL-SEIDEL [8])[1].

Setting $g(z) = 1/f(z)$ again, we have

**Corollary 3.** *There exists a non-constant function $g(z)$, meromorphic in $D: |z| < 1$, with the following property: There is a set of spirals $\sigma_t\,(0 \leqq t < 1)$ in $|z| < 1$, each of which approaches $\Gamma: |z| = 1$ asymptotically, such that every point of $|z| < 1$ belongs to one, and only one, of these spirals and, for every t,*

$$\lim_{\substack{|z|\to 1\\ z\in\sigma_t}} g(z) = 0\,.$$

Remark[2]. Corollary 3 is no longer valid if the set of spirals $\sigma_t\,(0 \leqq t < 1)$ is replaced by an analogous set of simple curves in $|z| < 1$, terminating at the points of $\Gamma$. Let $\lambda_0$ be a simple curve in $D$, starting from the origin and ending at $z = 1$, and denote by $\lambda_\theta$ the transform of $\lambda_0$ under the rotation $z' = e^{i\theta} z$. From Theorem 10, § 3 it follows that Corollary 3 is not valid, if we replace $\sigma_t\,(0 \leqq t < 1)$ by the family $\{\lambda_\theta\}$. To prove this, assuming the existence of such a function $g(z)$, we have only to show that for every $e^{i\theta_0} \in \Gamma$, $C_D(g, e^{i\theta_0})$ is total. If $C_D(g, e^{i\theta_0})$ is sub-total, then there exists a Fatou arc $\mathfrak{A}$ of $\Gamma$ containing $z_0 = e^{i\theta_0}$. Since $F(g)$ is metrically dense on $\mathfrak{A}$ and there are at most enumerably many ambiguous points of $g(z)$, $g(z)$ must be identically zero.

**6.** BAGEMIHL and SEIDEL [2—5] have obtained many important results on the boundary properties of analytic functions. Among these results, the following is the most important.

**Theorem 8.** *Let there be given a continuous complex-valued function $g(z)$ in $D: |z| < 1$ and a non-empty $F_\sigma$-set $E$ of category I on $\Gamma: |z| = 1$. Then, there exists a function $f(z)$, regular in $|z| < 1$, such that, for every $e^{i\theta} \in E$, the radial cluster set of $f(z)$, $\mathfrak{R}f(z)$, $\mathfrak{I}f(z)$ coincides with that of $g(z)$, $\mathfrak{R}g(z)$, $\mathfrak{I}g(z)$ respectively* (BAGEMIHL-SEIDEL [3])[3].

*Proof.* By hypothesis, there exists a decomposition of $E$:

$$E = \bigcup_{n=1}^{\infty} E_n\,, \tag{5}$$

---

[1] Each spiral $\sigma_t\,(0 \leqq t < 1)$ can be defined as the union of enumerably many circular arcs and enumerably many rectilinear segments in $|z| < 1$. For a related theorem, cf. BAGEMIHL-ERDÖS-SEIDEL [1], Theorem 4, p. 139.

[2] Cf. also BAGEMIHL-SEIDEL [8], p. 84.

[3] The original theorem, which generalizes results of ROTH [1], has been stated in a more general form, introducing a new notion "tress". See BAGEMIHL-SEIDEL [3], pp. 186—187.

where each $E_n$ is a closed nowhere dense set on $\Gamma (n = 1, 2, \ldots)$. We denote by $S_n$ the union of radii $\{z = re^{i\theta} : 0 \leqq r < 1, e^{i\theta} \in E_n\}$ $(n = 1, 2, \ldots)$, by $\Gamma_r$ the circumference $|z| = r$ $(0 < r < 1)$ and by $D_r$ the closed disc $|z| \leqq r$ $(0 < r < 1)$. Let $0 < r_0 < r_1 < \cdots < r_n < \cdots < 1$, where $\lim\limits_{n\to\infty} r_n = 1$, and set

$$F_0 = D_{r_0} \cup (S_1 \cap \Gamma_{r_1}),$$
$$F_n = D_{r_n} \cup [(\bigcup_{j=1}^{n} S_j) \cap (D_{r_{n+1}} - D_{r_n})] \cup (S_{n+1} \cap \Gamma_{r_{n+1}}),$$

$(n = 1, 2, \ldots)$.

Obviously, each $F_n$ is a bounded closed set whose complement relative to $D$ is connected. We now define, by induction, two real-valued functions $\varphi_n(z)$ and $\psi_n(z)$ on $F_n$, and a polynomial $p_n(z)$ in the following way. Put

$$\varphi_0(z) = \psi_0(z) = 0, \quad z \in D_{r_0},$$
$$\varphi_0(z) = \Re g(z), \quad \psi_0(z) = \Im g(z), \quad z \in S_1 \cap \Gamma_{r_1}.$$

The function $\varphi_0(z) + i\psi_0(z)$ is continuous on $F_0$ and regular at all interior points of $F_0$. Consequently, by a theorem of Mergelyan [1], there exists a polynomial $p_0(z)$ such that

$$|p_0(z) - [\varphi_0(z) + i\psi_0(z)]| \leqq 1, \quad z \in F_0.$$

Suppose, now, that we have defined the functions $\varphi_{n-1}(z)$ and $\psi_{n-1}(z)$ on $F_{n-1}$, and the polynomials $p_0(z), p_1(z), \ldots, p_{n-1}(z)$ such that

$$\varphi_{n-1}(z) = \Re g(z), \quad \psi_{n-1}(z) = \Im g(z), \quad z \in F_{n-1} \cap \Gamma_{r_n},$$

$$|p_0(z) + p_1(z) + \cdots + p_{n-1}(z) - [\varphi_{n-1}(z) + i\psi_{n-1}(z)]| \leqq 1/2^{n-1}, z \in F_{n-1};$$

hence

$$|p_0(z) + p_1(z) + \cdots + p_{n-1}(z) - g(z)| \leqq 1/2^{n-1}, \ z \in F_{n-1} \cap \Gamma_{r_n}. \tag{6}$$

By Tietze's theorem[1] and (6), we easily see that there are two real-valued functions $\xi_n(z)$ and $\eta_n(z)$ continuous in $|z| < 1$ such that

$$\begin{aligned} &\xi_n(z) = \Re \sum_{j=0}^{n-1} p_j(z), \ \eta_n(z) = \Im \sum_{j=0}^{n-1} p_j(z), \ z \in F_{n-1} \cap \Gamma_{r_n}; \\ &\xi_n(z) = \Re g(z), \ \eta_n(z) = \Im g(z), \ z \in F_n \cap \Gamma_{r_{n+1}}; \\ &|\xi_n(z) - \Re g(z)| \leqq 1/2^{n-1}, \ |\eta_n(z) - \Im g(z)| \leqq 1/2^{n-1}, \\ &\qquad z \in \bigcup_{j=1}^{n} S_j, \ r_n < |z| < r_{n+1}. \end{aligned} \tag{7}$$

We now put

$$\varphi_n(z) = \Re \sum_{j=0}^{n-1} p_j(z), \ \psi_n(z) = \Im \sum_{j=0}^{n-1} p_j(z), \ z \in D_{r_n}, \tag{8}$$

$$\varphi_n(z) = \xi_n(z), \ \psi_n(z) = \eta_n(z), \ z \in F_n \cap (D_{r_{n+1}} - D_{r_n}). \tag{9}$$

[1] See, e. g., Kuratowski [1], p. 117.

The function $\varphi_n(z) + i\psi_n(z)$ is evidently continuous on $F_n$ and regular at all interior points of $F_n$. Again by Mergelyan's theorem, there exists a polynomial $p_n(z)$ such that

$$|p_0(z) + p_1(z) + \cdots + p_n(z) - [\varphi_n(z) + i\psi_n(z)]| \leqq 1/2^n, \quad z \in F_n. \quad (10)$$

Let $f(z) = p_0(z) + p_1(z) + \cdots + p_n(z) + \cdots$ for $|z| < 1$. If $z \in D_{r_n}$, it follows from (8) and (10) that

$$|p_{n+j}(z)| \leqq 1/2^{n+j} \qquad (j = 0, 1, \ldots). \quad (11)$$

Consequently, $f(z)$ is regular in $|z| < 1$. It is easy to show that

$$\lim_{r\to 1} (\Re f(re^{i\theta}) - \Re g(re^{i\theta})) = 0$$

for every $e^{i\theta} \in E$. Suppose that $e^{i\theta} \in E_m$. Then, if $z = re^{i\theta}, r_{m+k} \leqq r < r_{m+k+1}$ $(k \geqq 0)$, we have

$$\begin{aligned} |\Re f(z) - \Re g(z)| &\leqq \left|\sum_{j=0}^{m+k} \Re p_j(z) - \varphi_{m+k}(z)\right| + |\varphi_{m+k}(z) - \Re g(z)| \\ &+ \left|\sum_{j=m+k+1}^{\infty} \Re p_j(z)\right| \qquad (12) \\ &\leqq \frac{1}{2^{m+k}} + \frac{1}{2^{m+k-1}} + \sum_{j=0}^{\infty} \frac{1}{2^{m+k+1+j}} \leqq \sum_{j=m+k-1}^{\infty} \frac{1}{2^j}. \end{aligned}$$

Similarly, for every $e^{i\theta} \in E$,

$$\lim_{r\to 1} (\Im f(re^{i\theta}) - \Im g(re^{i\theta})) = 0.$$

Consequently, we have, for every $e^{i\theta} \in E$,

$$\lim_{r\to 1} (f(re^{i\theta}) - g(re^{i\theta})) = 0.$$

Remark. If $E$ is closed and nowhere dense on $|z| = 1$, then we have $(f(re^{i\theta}) - g(re^{i\theta})) \to 0$ uniformly for $e^{i\theta} \in E$. For we can choose a fixed $m$ independent of $e^{i\theta}$ of $E$ in (12).

As an immediate consequence of Theorem 8, we have

**Theorem 9.** *There exists a function $f(z)$ regular in $|z| < 1$ such that for almost all $e^{i\theta}$ in the interval $0 \leqq \theta \leqq 2\pi$, $\Re f(re^{i\theta}) \to 0$ and $\Im f(re^{i\theta}) \to +\infty$ as $r \to 1$* (Bagemihl-Seidel [3]).

*Proof.* Let $E = \bigcup_n P_n$, where $P_n$ $(n = 1, 2, \ldots)$ is a perfect nowhere dense set of measure $2\pi - 1/n$, so that $E$ is an $F_\sigma$-set of category I and measure $2\pi$. Apply Theorem 8, taking $g(z) = i/(1 - |z|)$.

Setting $g(z) = e^{f(z)}$, we obtain

**Corollary 4.** *There exists a function $g(z)$ regular in $|z| < 1$ such that the radial cluster set $C_\varrho(f, e^{i\theta})$ coincides with the whole circumference $|w| = 1$ for almost every $e^{i\theta}$ of $\Gamma$: $|z| = 1$.*

By the same idea of the proof of Theorem 8, we can prove the following (Theorems 10—15).

**Theorem 10.** *Let $\mathcal{M}$ be a non-empty set on $\Gamma$: $|z| = 1$ and denote by $M$ the set of radii terminating in the points of $\mathcal{M}$. In order that there exists a function, regular in $|z| < 1$, which is uniformly bounded on $M$ and yet has no radial limit at any $e^{i\theta} \in \Gamma$, it is necessary and sufficient that $\mathcal{M}$ is nowhere dense* (BAGEMIHL-SEIDEL [3]).

*Proof.* The necessity follows from Fatou's theorem. To prove the sufficiency, let $E = \overline{\mathcal{M}}$ and $S$ be the set of all radii ending at the points of $E$. Let $r_n = 1 - 1/n$ $(n = 1, 2, \ldots)$. The complement of the set $S \cap \Gamma_{r_n}$ relative to $\Gamma_{r_n}$: $|z| = r_n$ consists of at most enumerably many open arcs. In each of these open arcs whose length $\lambda$ is greater than $1/n$, take a closed subarc of length $\lambda - 1/n$, and denote the union of these finitely many closed subarcs by $A_n$. Let $\{\alpha_n\}$ be a sequence of complex numbers dense in the $w$-plane. Take $g(z)$ to be a continuous function on $S \cup (\bigcup_{n=1}^{\infty} A_n)$ such that $g(z) = i/(1 - |z|)$ on $S$ and $g(z) = \alpha_n$ on $A_n$ $(n = 1, 2, \ldots)$. A simple modification of the proof of Theorem 8 yields a regular function $F(z)$ such that $\Re F(z) \to 0$ uniformly and $\Im F(z) \to +\infty$ as $|z| \to 1$ with $z \in S$ and $C_\varrho(f, e^{i\theta})$ is total at $e^{i\theta} \notin E$. The function $f(z) = e^{F(z)}$ has the required properties.

By a *monotonic boundary path,* we mean a simple continuous curve $z = z(t)$ $(0 \leqq t < 1)$ in $|z| < 1$, with $z(0) = 0$, such that, as $t \to 1$, $|z(t)| \to 1$ (strictly) monotonically.

**Theorem 11**[1]. *Let $\{\lambda_n\}$ be an enumerable number of monotonic boundary paths in $|z| < 1$, the intersection of every pair of which is the origin, and let $\{K_n\}$ be a sequence of arbitrary continua in the $w$-plane. Then, there exists a non-constant function regular in $|z| < 1$ such that for every $n$*

$$C_{\lambda_n}(f, e^{i\theta}) = K_n \qquad (n = 1, 2, \ldots)$$

(BAGEMIHL-SEIDEL [3]).

**Theorem 12.** *Let $\mathcal{M}_n$ $(n = 1, 2, \ldots)$ be enumerably many mutually exclusive subsets of $\Gamma$: $|z| = 1$ of respective positive measures $m_n$, and let $K_n$ $(n = 1, 2, \ldots)$ be arbitrary continua in the $w$-plane. Then, there exists a non-constant function $f(z)$ regular in $|z| < 1$ such that, for every $n$ and for almost every $e^{i\theta} \in \mathcal{M}_n$, the radial cluster set $C_\varrho(f, e^{i\theta})$ is $K_n$* (BAGEMIHL-SEIDEL [3])[2].

**Corollary 5.** *Let $K$ be an arbitrary continuum in the $w$-plane. Then, there exists a non-constant function $f(z)$ regular in $|z| < 1$ such that $C_\varrho(f, e^{i\theta}) = K$ for almost every $e^{i\theta}$ of $\Gamma$.*

[1] In case every $K_n$ is a single point, this result has been obtained in BAGEMIHL-SEIDEL [2].

[2] In the proof, the Hahn-Mazurkiewicz theorem (cf. C. KURATOWSKI [2], p. 188) is used.

**Theorem 13.** *Let $E$ be a non-empty closed nowhere dense set in the interval $0 \leqq \theta \leqq 2\pi$, and let $\alpha(\theta)$ and $\beta(\theta)$ be real-valued functions of Baire class 1 on $E$. Then, there exists a function $f(z)$, regular in $|z| < 1$, such that, for every $\theta \in E$, $\lim_{r \to 1} \Re f(re^{i\theta}) = \alpha(\theta)$, $\lim_{r \to 1} \Im f(re^{i\theta}) = \beta(\theta)$ and, for every $\theta \notin E$ in the interval $0 \leqq \theta \leqq 2\pi$, $C_\varrho(f, e^{i\theta})$ is total* (BAGEMIHL-SEIDEL [3])[1].

Let $f(z)$ be regular in $|z| < 1$. By the *radial limit set* of $f(z)$, we mean the set of values $\gamma$ such that for some $\theta$ in the interval $0 \leqq \theta \leqq 2\pi$, $\lim_{r \to 1} f(re^{i\theta}) = \gamma$.

**Theorem 14**[2]. *A necessary and sufficient condition that a set $A$ in the extended $w$-plane be the radial limit set of a function regular in $|z| < 1$ is that $A$ be an analytic set* (BAGEMIHL-SEIDEL [3]).

Applying Theorem 13, LEHTO has proved

**Theorem 15.** *Let $\psi_1(\theta)$ and $\psi_2(\theta)$ be arbitrary measurable functions in the interval $0 \leqq \theta \leqq 2\pi$. Then, there exists a function $f(z)$, regular in $|z| < 1$, such that*

$$f(e^{i\theta}) = \psi_1(\theta) + i\,\psi_2(\theta)$$

*almost everywhere, where $f(e^{i\theta})$ denotes the radial limit of $f(z)$ at $e^{i\theta}$* (LEHTO [8]).

Remark[3]. Let $\psi(\theta)$ be measurable and finite almost everywhere for $0 \leqq \theta \leqq 2\pi$. Without making use of Theorem 15, LEHTO has constructed an analytic expression for harmonic function $u(z)$ for which $u(e^{i\theta}) = \psi(\theta)$ almost everywhere. Lehto's method is based on the following lemma: Let $\psi(\theta)$ be a measurable function which is finite almost everywhere, and let $F(z)$ be a regular function in $|z| < 1$ such that $|\Re F(e^{i\theta})| < \infty$, $\Im F(e^{i\theta}) = 0$ for almost all $\theta$, and $\psi(\theta)/F(e^{i\theta})$ is integrable in Lebesgue's sense. Then

$$u(z) = \frac{1}{2\pi} \Re \left\{ F(z) \int_0^{2\pi} \frac{\psi(\theta)}{F(e^{i\theta})} \frac{e^{i\theta} + z}{e^{i\theta} - z}\, d\theta \right\}$$

is a harmonic function in $|z| < 1$ such that $u(e^{i\theta}) = \psi(\theta)$ almost everywhere (LEHTO [8]; cf. also LOKKI [1]).

**7.** We state briefly some results of different type on boundary behaviours. Let $f(z)$ be regular in the unit disc $D: |z| < 1$. Let $\Gamma$ denote the unit circumference $|z| = 1$. We say that $e^{i\theta}$ is a Lusin *point* for $f(z)$ provided that $f(z)$ maps every circular disc internally tangent to $\Gamma$ at $e^{i\theta}$ onto

---

[1] For the extension of this theorem to the case of a multiply connected domain, see BAGEMIHL-SEIDEL [5].

[2] For the extension of this theorem, see BAGEMIHL-SEIDEL [4].

[3] In this remark, $u(e^{i\theta})$ and $F(e^{i\theta})$ denote radial limits of $u(z)$ and $F(z)$ at $e^{i\theta}$ respectively.

a Riemann domain of infinite area. LOHWATER and PIRANIAN [3] proved that *there exists a function* $f(z)$ *of the form* $\Sigma a_k z^k$ *with* $\Sigma |a_k| < \infty$ *for which every point* $e^{i\theta}$ *is a Lusin point*[1]. From a general theorem of PIRANIAN [1] follows that *if* $E$ *is a set of type* $G_\delta$ *on* $\Gamma$, *then there exists a function* $f(z)$ *which is regular in* $D$ *and continuous in* $\overline{D}$, *and for which every point of* $E$ *is a Lusin point and no point of* $\Gamma - E$ *is a Lusin point.* Furthermore, PIRANIAN [1] proved that *if* $E$ *is a set of type* $G_\delta$ *on* $\Gamma$, *then there exists a function* $f(z)$ *which is regular in* $D$, *and which maps the radius of* $e^{i\theta}$ *onto a curve of infinite (Euclidean or spherical) length if and only if* $e^{i\theta} \in E$ [2].

## § 5. Meromorphic functions of bounded type and normal meromorphic functions

Recently LEHTO [7] has studied systematically the value distribution of a meromorphic function of bounded type, introducing a new method based on the subharmonicity of the counting function $N(r, a)$ of the variable $a$. More recently, LEHTO and VIRTANEN [1], [2] have introduced normal meromorphic functions and studied their boundary behaviours, by applying Montel's theory of normal families and the theory of hyperbolic measure. The main purpose of this section is to state some interesting results of LEHTO and LEHTO-VIRTANEN from the view-point of cluster sets.

1. Let $w = f(z)$ be non-constant and meromorphic in the unit circle $|z| < 1$. Let $n(r, a)$ denote the number of $a$-points, counted according to their multiplicity, in the disc $|z| \leqq r\ (< 1)$. Consider the counting function

$$N(r, a) = \int_0^r \frac{n(r, a)}{r}\, dr\,.$$

By partial integration, we have

$$N(r, a) = \sum_{|z_\nu| < r} \log \frac{r}{|z_\nu|}\,, \tag{1}$$

where $z_1, z_2, \ldots$ denote the $a$-points of $f(z)$. Since Green's function is conformally invariant, $N(r, a)$ can be written in the form

$$N(r, a) = \sum_\nu g(P_\nu(a), Q, F_r)\,, \tag{2}$$

where $F_r$ denotes the Riemann (covering) surface onto which $w = f(z)$ maps the disc $|z| < r$, $P_\nu(a)$ $(\nu = 1, 2, \ldots)$ are all points of $F_r$ above $w = a$, $Q$ is the image of $z = 0$ on $F_r$, and $g(P, Q, F_r)$ is the Green's function of $F_r$ with pole at $Q$.

From the representation (2), it follows that for a fixed $r$, $N(r, a)$ is continuous in $a$. Moreover, using the well-known mean-value theorem, we

[1] As to its related results and extensions, see JENKINS [1], PIRANIAN and RUDIN [1], PIRANIAN and SHIELDS [1], BAGEMIHL [4], PIRANIAN [1].

[2] As to its related results and extensions, see HERZOG and PIRANIAN [1], RUDIN [4], BAGEMIHL [5].

can easily prove that $N(r, a)$ is subharmonic in $a$, except for the logarithmic pole at $a = f(0)$ (for details, see LEHTO [1], p. 7).

Applying the argument principle to $f(z)$, we obtain

$$\frac{1}{2\pi}\int_0^{2\pi} d\arg\left(\frac{f(re^{i\theta})-a}{f(re^{i\theta})-\zeta}\right) = n(r, a) - n(r, \zeta)\,. \tag{3}$$

Let $\mu$ be a completely additive set function defined for the Borel subsets of an arbitrary closed set $S$. Set

$$u(w) = \int_S \log|w-\zeta|\, d\mu(\zeta) - \mu(S)\log|w-a|$$

and suppose that $u(w)$ is continuous at $w = f(0)$. By integrating (3) first with respect to $\mu(\zeta)$ and then with respect to $r$ after dividing (3) by $r$, we obtain, for $a \neq f(0)$, the relation

$$\frac{1}{2\pi}\int_0^{2\pi} u(f(re^{i\theta}))\, d\theta - u(f(0)) + N(r, a)\,\mu(S) = \int_S N(r, \zeta)\, d\mu(\zeta)\,. \tag{4}$$

Now, let $G$ be a domain in the $w$-plane with boundary $B$ of positive capacity. We write for $a$ in $G$

$$g^+(a, w, G) = \begin{cases} g(a, w, G), & w \in G \\ 0, & w \notin G\,, \end{cases}$$

where $g(a, w, G)$ is the Green's function of $G$ with singularity at $a = w$. Let $\omega(a, e, G)$ be the value of the harmonic measure of the set $e\,(\subset B)$ at the point $w = a$ with respect to $G$. If we take in (4) as $S$ the boundary $B$ and put $\mu(e) = \omega(a, e, G)$, then (4) becomes

$$\Phi(r, a, G) + N(r, a) = P(r, a, G)\,, \tag{5}$$

where

$$\Phi(r, a, G) = \frac{1}{2\pi}\int_0^{2\pi} g^+(f(re^{i\theta}), a, G)\, d\theta$$

and

$$P(r, a, G) = \int_B N(r, \zeta)\, d\omega(a, \zeta, G) + g^+(a, f(0), G)\,. \tag{6}$$

Evidently, $P(r, a, G)$ is the least harmonic majorant of $N(r, a)$ in $G$.

**2.** We now suppose that $f(z)$ is of bounded type. Obviously, $N(a) = \lim\limits_{r\to 1} N(r, a)$ is finite, except at the point $a = f(0)$. By Harnack's theorem, $P(r, a, G)$ converges to a harmonic limiting function $P(a, G)$ for $r \to 1$ and we can write

$$P(a, G) = \int_B N(\zeta)\, d\omega(a, \zeta, G) + g^+(a, f(0), G)\,. \tag{7}$$

By definition, $P(a, G)$ is the least harmonic majorant of $N(a)$ in $G$[1]. Consequently, if we write

$$\Phi(a, G) = \lim_{r \to 1} \Phi(r, a, G) ,$$

then the relation (5) becomes

$$\Phi(a, G) + N(a) = P(a, G) . \tag{8}$$

It is well-known that $f(z)$ has a radial limit almost everywhere on $|z| = 1$. Denote by $H$ the closure of the set of the radial limits of $f(z)$ which corresponds to a set on $|z| = 1$ of measure $2\pi$. We now suppose that $H$ is sub-total; i. e., the complement $\mathscr{C}H$ with respect to the $w$-plane is not empty[2]. Select an arbitrary domain $G$ in the exterior of $H$. Under these hypotheses, LEHTO has obtained, again with an aid of the general relation (3) by choosing $\mu$ as the difference of two suitable harmonic measures, the following beautiful

**Theorem 1.** *The "Schmiegungsfunktion"*

$$\Phi(a, G) = \frac{1}{2\pi} \lim_{r \to 1} \int_0^{2\pi} g^+(f(re^{i\theta}), a, G)\, d\theta$$

*vanishes, except perhaps for a set of values a of capacity zero. In other words, up to such a null-set, $N(a)$ coincides with its least harmonic majorant $P(a, G)$ in $G$* (LEHTO [7])[3].

From this theorem, it follows that $f(z)$ *either takes no value in $G$, or it takes all values there with the possible exception of the above null-set* (LEHTO [5])[4].

Suppose that $w = f(0)$ lies in $G$. Then, by (7), we have $P(a, G) \geqq g(a, f(0), G)$, and Theorem 1 yields the inequality (LEHTO [5])

$$N(a) \geqq g(a, f(0), G) , \tag{9}$$

which is valid, up to a set of capacity zero.

[1] For a fixed $r$, $N(r, a)$ is continuous and subharmonic in $a$ except for the logarithmic pole $a = f(0)$. Consequently, as a limit of monotonic increasing continuous functions, $N(a)$ is lower semi-continuous but has the following characteristic property of subharmonic functions: Let $G$ be an arbitrary domain with boundary $B$ such that $\bar{G}$ does not contain $a = f(0)$, and let $U(a)$ be a function harmonic in $G$ and continuous in $\bar{G}$, and $N(a) \leqq U(a)$ on $B$. Then we also have $N(a) \leqq U(a)$ in $G$. We should recall that in the usual definition of subharmonic functions the upper semi-continuity is assumed.

[2] There exists a function of bounded type in $|z| < 1$ which admits every complex value as an angular limit in a null-set on $|z| = 1$ (LEHTO [7], pp. 42—43).

[3] For the proof, see LEHTO [7], pp. 12—17.

[4] Since $P(a, G)$ is non-negative, $P(a, G)$ cannot have zeros in $G$ without vanishing identically. Hence, if $w = f(z)$ takes one value $a$ belonging to $G$, i. e. if $N(w) > 0$ at the point $w = a$, we have $P(a, G) > 0$ in the whole domain $G$. Hence, $N(a) > 0$, except perhaps for a set of capacity zero; in other words, up to such a set, $f(z)$ assumes every value belonging to $G$.

If the value set of $f(z)$ in $|z| < 1$ is contained in $G$, then the function $N(a)$ vanishes identically on the boundary $B$ of $G$. Hence, by (7)

$$P(a, G) = g(a, f(0), G);^{1} \tag{10}$$

hence (8) becomes

$$\Phi(a, G) + N(a) = g(a, f(0), G) \tag{11}$$

which shows an analogy with the first main theorem in the theory of meromorphic functions. By Theorem 1, in this case, we have

$$N(a) = g(a, f(0), G) \tag{12}$$

with a possible exception of a set of capacity zero.

3. Let $w = f(z)$ be a function of bounded type in the unit circle. Denote by $H$ the closure of the set of the radial limits of $f(z)$ which corresponds to a set on $|z| = 1$ of measure $2\pi$. We suppose that $H$ is sub-total. Take the domain $G$ in the exterior of $H$ which contains at least one value $w = f(z)$. LEHTO introduced the new definition of *deficiency* $\delta(a)$ of $f(z)$ with respect to $a$ $(\neq f(0))$ belonging to $G$:

$$\delta(a) = 1 - N(a)/P(a, G); \tag{13}$$

this definition is formally equivalent to the classical one. We call a value $a$ *normal* for $f(z)$ if $\delta(a) = 0$, *deficient* if $\delta(a) > 0$. By Theorem 1, *all the values of G are normal, except for at most a set of capacity zero.* We note that $\delta(a)$ is independent of the choice of the domain $G$. In fact, if $G'$ is a sub-domain of $G$, then $P(a, G) = P(a, G')$ at every point of $G'$. This follows immediately from the fact that both harmonic functions $P(a, G)$ and $P(a, G')$ coincide with $N(a)$ except for a set of capacity zero[2].

Now, we consider a covering surface $F$, of the $w$-plane, generated by $w = f(z)$. Denote by $F_\nu$ $(\nu = 1, 2, \ldots)$ the distinct connected parts of $F$ above the domain $G$. Let $F_\nu^\infty$ be the universal covering surface of $F_\nu$. Denote by $w = \psi_\nu(z)$ the function mapping $|z| < 1$ conformally onto $F_\nu^\infty$. Then, it is easily seen that (i) all the values of $w = \psi_\nu(z)$ lie in $G$ and (ii) the radial limits $\psi_\nu(e^{i\vartheta})$ belong to the boundary of $G$ for almost all $\vartheta$ [3]. By (12), we have

$$N(a, \psi_\nu) = g(a, \psi_\nu(0), G), \tag{14}$$

except perhaps for a set of capacity zero.

---

[1] This case is particularly interested, because $P(a, G)$ is a domain function and is independent of $f(z)$.

[2] Compared with the classical deficiency $1 - \overline{\lim_{r\to 1}} \frac{N(r, a)}{T(r)}$, $T(r)$ being Nevanlinna's characteristic function of $f(z)$, the deficiency (13) in the sense of LEHTO has the advantage of being independent of metrical quantity $r$.

[3] To prove this fact, we have only to use the idea in the proof of Theorem 4, § 1. For details see LEHTO [7], pp. 24—26. In the special case where $G$ is the unit circle $|w| < 1$, evidently $\psi_\nu(z)$ is a function of class $(U)$.

LEHTO has proved

**Theorem 2.** (Principle of localization.) *A value a outside H is normal for f(z) if and only if it is normal for every function* $\psi_k(z)$, $k = 1, 2, \ldots$ *Here* $\psi_k(z)$ *denote all the functions which map the unit circle onto the universal covering surfaces of the connected parts of F which lie above an arbitrary fixed domain G containing the point a and located outside H* (LEHTO [7]).

Applying this theorem and a previous result of LEHTO [3] (Theorem 4, p. 10), we can prove

**Theorem 3.** *Let* $z' = L(z)$ *be a linear transformation of the unit circle onto itself. Then, a is a normal value for* $f(L(z))$ *if and only if a is normal for* $f(z)$ (LEHTO [7]).

According to Theorem 2, we can now apply the theory of functions of class $(U)$, by taking as domain $G$ a circular disc $\Delta: |w-a| < \varrho$. Using the canonical representation (see Theorem 1, § 1),

$$\psi_k(z) = a + B(z)\, h(z)\,,$$

where $B(z)$ is the Blaschke product extended over the zeros $z_\nu$ $(\nu = 1, 2, \ldots)$ of $\psi_k(z) - a$ and

$$h(z) = \varrho e^{i\gamma} \exp\left[-\frac{1}{2\pi}\int_0^{2\pi} \frac{e^{i\theta}+z}{e^{i\theta}-z}\, d\mu(\theta)\right],$$

$\mu(\theta)$ being a monotonic non-decreasing function of $\theta$ and $\gamma$ a real constant. We have

$$|\psi_k(0) - a| = |h(0)| \prod_\nu |z_\nu|\,.$$

Since $|h(0)| \leqq \varrho$, it follows that

$$|\psi_k(0) - a| \leqq \varrho \prod_\nu |z_\nu|\,. \tag{15}$$

The equality holds if and only if $|h(0)| = \varrho$; in this case $h(z) \equiv \varrho e^{i\gamma}$. Now $\log \frac{\varrho}{|w-a|}$ is the Green's function of the circle $|w-a| < \varrho$ with pole at $w = a$. Further $-\log \prod_\nu |z_\nu| = N(a, \psi_k)$. Hence, (15) is equivalent to the inequality

$$N(a, \psi_k) \leqq g(a, \psi_k(0), \Delta)\,.$$

Thus, it follows that $a$ is a normal value for $\psi_k(z)$ if and only if $h(z)$ is constant. By Theorem 2, § 1, we see that if a value in $\Delta$ is deficient for $\psi_k(z)$, then it is always a radial (angular) limit. Consequently, in virtue of Theorem 2 in this section, we obtain the following beautiful

**Theorem 4.** *Let* $w = a$ *be a point outside H, and let a be deficient for* $f(z)$. *Then, a is an asymptotic value of* $f(z)$ *with its asymptotic path terminating at a point of* $|z| = 1$ (LEHTO [7]).

Remark. The corresponding results do not hold for meromorphic functions with unbounded characteristic[1].

Remark. Frostman's example (see, Remark of Theorem 2, § 1) shows that an asymptotic value is not necessarily deficient. If we consider direct and indirect critical singularities of the covering surface generated by $w = f(z)$, then direct critical singularities always correspond to deficient values, whereas this need not be the case for indirect critical singularities (e. g. Frostman's example stated above)[2].

**4.** LEHTO and VIRTANEN [1] have introduced the following definition: A meromorphic function $f(z)$ is called *normal* in a simply connected domain $D$, if the family $\{f(S(z))\}$ is normal, where $z' = S(z)$ denotes an arbitrary conformal mapping of $D$ onto itself[3].

We first state the following

**Theorem 5.** *Let $f(z)$ be meromorphic in $D: |z| < 1$. Suppose that $f(z)$ has the asymptotic value zero along a Jordan curve lying in D and terminating at a point $z_0$ on $|z| = 1$ and that $f(z)$ does not possess the angular limit zero at the point $z_0$. Then, there exist a Jordan curve L in D with end-point at $z_0$, on which $f(z)$ tends to zero, and a sequence of points $\{z_n\}$, in D, which converges to $z_0$ and at which $f(z_n) = a$ for a certain value $a \neq 0$, such that the points $z_n$ have a bounded hyperbolic distance from L less than M, where M is any fixed positive number* (LEHTO-VIRTANEN [1])[4].

As an immediate consequence of the above theorem, we have

**Theorem 6.** *Let $f(z)$ be meromorphic and normal in $D: |z| < 1$ and let $f(z)$ have an asymptotic value $\alpha$ at a point $z_0$ on $|z| = 1$ along a Jordan curve lying in D. Then, $f(z)$ possesses the angular limit $\alpha$ at the point $z_0$* (LEHTO-VIRTANEN [1]).

*Proof*[5]. Without loss of generality, we may suppose that $\alpha = 0$. Contrary to the assertion, suppose that $f(z)$ does not possess the angular limit zero at $z_0$. Consider now the asymptotic path $L$ and the sequence of points $\{z_n\}$ in Theorem 5. Denote by $z' = S_n(z)$ a function which maps $D$ conformally onto itself and satisfies the condition $S_n(0) = z_n$. We denote by $K$ the hyperbolic circle with center at $z = 0$ and of hyperbolic radius $M + 1$. By hypothesis, the family $\{f(S_n(z))\}$ is normal in $D$; hence, we can select a subsequence $\{f(S_{n_k}(z))\}$ which converges uniformly on every

[1] As is well-known, Mme. SCHWARTZ, TEICHMÜLLER and HAYMAN have given counterexamples of meromorphic functions for which an exceptional value (in the sense of NEVANLINNA) is not asymptotic (cf. WITTICH [1]).

[2] LETHO [7], Theorem 9, p. 41. HUCKEMANN [1] has shown that indirect critical singularities may give rise to deficient values.

[3] In the case of the unit disc, the writer said such a function to belong to class $(A)$ and obtained some results (see, NOSHIRO [3]).

[4] For the proof, see LEHTO-VIRTANEN [1], Theorem 1, pp. 49—52.

[5] An entirely similar argument was used in NOSHIRO [3]. See also recent related results of SEIDEL [3].

compact subset of $D$ to a meromorphic function $\varphi(z)$ in $D$. Since the images of the arcs of $L$ mapped by the inverse functions $z = S_{n_k}^{-1}(z')$ into the interior of $K$ has at least one accumulation continuum $\gamma$, it follows that $\varphi(z) = 0$ on $\gamma$ and therefore $\varphi(z)$ vanishes identically. However, this is a contradiction, because $f(S_{n_k}(0)) = a \neq 0$.

We shall now state the conditions for $f(z)$ to be normal. For this purpose, we introduce the spherical derivative

$$\varrho(f(z)) = \frac{|f'(z)|}{1 + |f(z)|^2}$$

of $f(z)$. It is well-known[1] that a family $\mathfrak{F}$ of meromorphic functions is normal in a domain $G$ if and only if there exists a positive constant $M$ such that

$$\varrho(f(z)) < M \text{ for all } f(z) \in \mathfrak{F} \tag{16}$$

on every compact set in $G$.

Using this criterion, we get

**Theorem 7.** *A non-constant function $f(z)$, meromorphic in $D$: $|z| < 1$, is normal if and only if*

$$\varrho(f(z))\,|dz| \leqq C\, d\sigma(z) \tag{17}$$

*is satisfied at every point of $D$, where $d\sigma(z) = \frac{|dz|}{1-|z|^2}$ and $C$ is a fixed finite positive constant* (NOSHIRO [3], LEHTO-VIRTANEN [1]).

*Proof.* Let

$$z' = S(z) = e^{i\alpha}\frac{z+\zeta}{1+\bar{\zeta}z} \quad (\alpha \text{ real}, \ |\zeta| < 1)\,.$$

Since $d\sigma(z)$ is conformally invariant, we have

$$(1-|z|^2)\,\varrho(f(S(z))) = (1-|z'|^2)\,\varrho(f(z'))\,; \tag{18}$$

hence, if (17) is satisfied at every point of $D$, then $f(z)$ is normal in $D$, by the criterion stated above. Next suppose that $f(z)$ is normal in $D$, i. e., the family $\{f(S(z))\}$ is normal in $D$. Then, by (16) $\sup\limits_S \varrho(f(S(0))) = C < +\infty$. Substituting $z = 0$ and $z' = e^{i\alpha}\zeta$ in (18), we have

$$\varrho(f(S(0))) = (1-|\zeta|^2)\,\varrho(f(e^{i\alpha}\zeta))\,,$$

whence follows (17).

Remark. From the proof, it is easily seen that a non-constant function, meromorphic in $|z| < 1$, is normal if and only if $\varrho(f(S(0)))$ is bounded for all $S$.

Applying this remark, we prove

**Theorem 8.** *Let $f(z)$ be meromorphic in $D$: $|z| < 1$. Let $A(r,f)$ denote the spherical area of the Riemannian image of the disc $|z| < r$ and $L(r,f)$*

---

[1] AHLFORS [7], p. 169. Cf. also MARTY [1].

*denote the spherical length of the image of the circumference $|z| = r$. Suppose that*

$$A(r, f(S(z))) \leqq k\, L(r, f(S(z))) \quad \text{for } 0 < r < 1\,, \tag{19}$$

*where $z' = S(z)$ denotes an arbitrary conformal mapping of $D$ onto itself and $k$ is a fixed constant independent of $S$ and of $r$. Then $f(z)$ is normal in $D$.*

*Proof.* It is only necessary to prove that $\varrho(f(S(0)))$ is bounded for all $S$. Put $\varrho(f(S(0))) = \alpha$ for a fixed $S$. The case of $\alpha = 0$ is trivial. Suppose that $\alpha > 0$ and put $z = \zeta/\alpha$. Then $g(\zeta) = f(S(\zeta/\alpha))$ is meromorphic in $|\zeta| < \alpha$. Evidently, $A(r, g) \leqq kL(r, g)$ for $0 < r < \alpha$. Moreover, $\varrho(g(0)) = \left|\frac{1}{\alpha}\right| \varrho(f(0)) = 1$. Assume, for a moment, that $\alpha > 1$. Applying a well-known inequality,

$$[L(r, g)]^2 \leqq 2\pi r \frac{dA(r, g)}{dr}$$

and (19), we have

$$\frac{dr}{r} \leqq 2\pi k^2 \frac{dA(r, g)}{[A(r, g)]^2}\,,$$

whence follows

$$\int_1^{r_0} \frac{dr}{r} \leqq 2\pi k^2 \left[\frac{1}{A(1, g)} - \frac{1}{A(r_0, g)}\right] (1 < r_0 < \alpha)$$

$$\leqq 2\pi k^2/A(1, g) \leqq 2\pi k^2/B\,,$$

where $B$ is an absolute constant[1].

Hence

$$\alpha \leqq e^{\frac{2\pi k^2}{B}}. \tag{20}$$

As the right hand side is greater than 1, (20) remains valid, without the restriction $\alpha > 1$.

Remark. Obviously the condition (19) is closely related to the fact that the covering surface generated by $w = f(z)$ is not regularly exhaustible in the sense of AHLFORS [2].

Applying Ahlfors' theory of covering surfaces[2] and using Theorem 8, we obtain several sufficient conditions for $f(z)$ to be normal. The following is typical.

**Theorem 9.** *Let $w = f(z)$ be meromorphic in $|z| < 1$. Let $\Delta_1, \Delta_2, \ldots, \Delta_q$ ($q \geqq 3$) be $q$ mutually disjoint closed Jordan domains on the Riemann*

[1] Here we use an extension of Bloch's theorem which states: Let $w = f(z)$ be meromorphic in $|z| \leqq 1$ and $\varrho(f(0)) = 1$. Then, the covering surface generated by $w = f(z)$ for $|z| < 1$ contains a schlicht spherical disc of a fixed radius $\mathfrak{B}$. We take as $B$ the area of the disc. This extension is due to AHLFORS [1]. Modifying Landau's well-known method in proving the classical Bloch's theorem, TSUJI [2, 12, 17] gave an elementary proof for this extension.

[2] AHLFORS [2].

*w-sphere. Denote by $\mu_j$ ($j = 1, 2, \ldots, q$) the minimum of the numbers of sheets of islands of F above $\Delta_j$, where F denotes the covering surface generated by $w = f(z)$* [1]. *Suppose that*

$$\sum_{j=1}^{q}\left(1-\frac{1}{\mu_j}\right) > 2 .$$

*Then, $f(z)$ is normal in $|z| < 1$.*

*Proof.* For simplicity, we put $g(z) = f(S(z))$ for a fixed $S$. Denote by $F_r$ the covering surface generated by $w = g(z)$ for $|z| < r$ ($< 1$) and by $\mu_j(r)$ the minimum of numbers of sheets of islands of $F_r$ above $\Delta_j$. Then $\sum_{j=1}^{q}(1-1/\mu_j(r))$ is a decreasing function of $r$ for $0 < r < 1$. Therefore,

$$\sum_{j=1}^{q}(1-1/\mu_j(r)) \geqq \sum_{j=1}^{q}(1-1/\mu_j) \geqq 2 + \frac{1}{42} .$$

On the other hand, by Ahlfors' theorem,

$$\sum_{j=1}^{q}\left(1-\frac{1}{\mu_j(r)}\right) \leqq 2 + h\,\frac{L(r,g)}{A(r,g)} ,$$

where $h$ ($> 0$) depends only upon $\Delta_j$. Hence,

$$A(r,g) \leqq 42hL(r,g) \quad \text{for } 0 < r < 1 .$$

Consequently, $f(z)$ is normal in $|z| < 1$, by the preceding theorem[2].

**Corollary.** *A function meromorphic in $|z| < 1$ is normal, if one of the following conditions is satisfied:*

(i) *$f(z)$ omits three different values in $|z| < 1$,*

(ii) *the covering surface F has no schlicht island above five mutually disjoint Jordan closed domains $\Delta_1, \Delta_2, \ldots, \Delta_5$ on the Riemann w-sphere.*

**5.** It may be of interest to compare meromorphic functions of bounded type with normal meromorphic functions. First, note that a meromorphic function of bounded type can have asymptotic values which are not angular limits: An example of Lehto [7]

$$f(z) = B(z)\,e^{\frac{1+z}{1-z}} , \quad B(z) = \prod_{k=2}^{\infty} \frac{\left(1-\frac{1}{k^2}\right) - z}{1-\left(1-\frac{1}{k^2}\right)z}$$

has the asymptotic value $w = \infty$ at $z = 1$ but this is not an angular limit (see also an example of Bagemihl-Seidel [6]). On the other hand, if a normal meromorphic function in $|z| < 1$ has an asymptotic value at a point $z_0$ of $|z| = 1$, then this is necessarily an angular limit at $z_0$ (Theorem 6).

[1] If there is no island of $F$ above $\Delta_j$, then we put $\mu_j = +\infty$.

[2] By a similar argument, we can prove that if $F$ is a covering surface of a closed Riemann surface of genus $\geqq 2$, $f(z)$ is normal in $|z| < 1$. Theorem 9 is essentially the same as a theorem of Dufresnoy [1], p. 223. As for related theorems, see Yûjôbô [3], Tsuji [12].

It is well-known that a meromorphic function of bounded type has an angular limit at almost every point of $|z| = 1$. However, as is easily seen from Theorem 9, *there exists a normal meromorphic function which possesses no asymptotic value*[1]. If $f(z)$ is normal, then, by Theorem 7,

$$\varrho(f(z)) \leqq \frac{C}{1-|z|^2} \quad \text{for } |z| < 1$$

and hence

$$T(r) = \int_0^r \frac{dr}{r} \int\int_{|z|<r} \varrho(f(z))\, dx\, dy = O\left(\log \frac{1}{1-r}\right).$$

# IV. Conformal mapping of Riemann surfaces

## § 1. Gross' property of covering surfaces

**1. Riemann surfaces of parabolic type.** Throughout this chapter, we shall consider Riemann surfaces in the sense of WEYL and RADÓ. Let $F$ be an arbitrary open Riemann surface of finite or infinite genus and $\{F_n\}$ $(n = 0, 1, \ldots)$ be a sequence of (relatively) compact domains of $F$ which satisfies:

(i) the boundary $\Gamma_n$ of $F_n$ consists of a finite number of simple closed analytic curves,

(ii) $\overline{F}_n \subset F_{n+1}$ $(n = 0, 1, \ldots)$ where $\overline{F}_n$ denotes the closure of $F_n$,

(iii) every connected component of the open set $F - \overline{F}_n$ consists of a finite number of non-compact domains,

(iv) $F = \bigcup_{n=0}^{\infty} F_n$.

Then, the sequence $\{F_n\}$ is called an *exhaustion* of $F$.

Consider the open set $F_n - \overline{F}_0$ which consists of a finite number of connected components and the harmonic measure

$$\omega(p, \Gamma_n, F_n - \overline{F}_0) \quad (n = 1, 2, \ldots) \tag{1}$$

in $F_n - \overline{F}_0$ with boundary values 1 on $\Gamma_n$ and 0 on $\Gamma_0$. By the maximum principle, the sequence $\omega(p, \Gamma_n, F_n - \overline{F}_0)$ is monotonically decreasing and converges uniformly to a limiting function $\omega(p)$ on any compact set of $F - \overline{F}_0$. Denoting by $\Gamma$ the ideal boundary of $F$, we write

$$\omega(p) = \omega(p, \Gamma, F - \overline{F}_0).$$

Evidently $0 \leqq \omega(p) < 1$ in $F - \overline{F}_0$. There arise two cases according as $\omega(p) > 0$ or $\omega(p) \equiv 0$. This distinction is independent of the choice of exhaustion $\{F_n\}$. We say that $F$ is *of parabolic type* or *of hyperbolic type* according as $\omega(p) \equiv 0$ or $\omega(p) > 0$. Let $D_n$ be the Dirichlet integral of $\omega(p, \Gamma_n, F_n - \overline{F}_0)$ on $F_n - \overline{F}_0$. Then, *for $F$ to be of parabolic type, it is*

[1] For example, some Schwarz's triangle functions. See also NOSHIRO [3], p. 154.

*necessary and sufficient that the monotone decreasing sequence* $\{D_n\}$ *converges to zero* (NEVANLINNA [2]). Let $p_0$ be a fixed point of $F_0$ and $g_n = g(p, p_0, F_n)$ be Green's function of $F_n$ with pole $p_0$. By Harnack's theorem, the monotone increasing sequence $\{g_n\}$ converges uniformly on any compact set of $F$ to Green's function $g(p, p_0, F)$ of $F$ or to a constant $+\infty$. We say that there exists no Green's function on $F$ in the latter case and that $F$ belongs to $O_G$. It is proved that *$F$ is of parabolic type if and only if $F \in O_G$* (P. J. MYRBERG [2], PARREAU [1], SARIO [9]). It is also known that $F$ *belongs to* $O_G$ *if and only if there exists no non-constant negative subharmonic function on* $F$ (OHTSUKA [2], AHLFORS [6]).

**2. Harmonic modulus.** Let $G$ be a relatively compact open set (not necessarily connected) on an open Riemann surface $F$. We suppose that the boundary $\gamma$ of $G$ consists of a finite number of closed analytic curves. We divide the boundary $\gamma$ into two disjoint sets $\gamma_0$ and $\gamma_1$. Let $\omega$ be the harmonic measure in $G$ with boundary values 0 on $\gamma_0$ and 1 on $\gamma_1$, and $\bar{\omega}$ be its conjugate harmonic function. We denote by $d$ the total variation of $\bar{\omega}$ on $\gamma_0$, i. e., $\int_{\gamma_0} d\bar{\omega} = d$, where the sense of $\gamma_0$ is negative with respect to $G$. We now put

$$z = z(p) = \frac{2\pi}{d}(\omega + i\bar{\omega}) = x + iy, \tag{2}$$

then $x = \frac{2\pi}{d}\omega$ is a function harmonic in $G$, which assumes 0 on $\gamma_0$ and $\frac{2\pi}{d}$ on $\gamma_1$, and the total variation of $y = \frac{2\pi}{d}\bar{\omega}$ on $\gamma_0$ is equal to $2\pi$. We call $\mu = 2\pi/d$ the *harmonic modulus* or simply the *modulus* of $G$ (PFLUGER [2])[1].

**3. Graph of a Riemann surface.** Let $F$ be an arbitrary open Riemann surface of finite or infinite genus and $\{F_n\}$ $(n = 0, 1, \ldots)$ be an exhaustion of $F$. Then, the open set $F_n - \bar{F}_{n-1}$ consists of a finite number of connected components $F_n^{(1)}, F_n^{(2)}, \ldots, F_n^{(i)}$ $(i = i(n))$. Evidently the boundary of $F_n - \bar{F}_{n-1}$ consists of $\Gamma_{n-1}$ and $\Gamma_n$. Denote by $\omega_n$ the harmonic measure in $F_n - \bar{F}_{n-1}$ with boundary values 1 on $\Gamma_n$ and 0 on $\Gamma_{n-1}$, and by $\bar{\omega}_n$ its conjugate. We denote by $d_n$ the total variation of $\bar{\omega}_n$ on $\Gamma_{n-1}$ and by $d_n^{(\nu)}$ the total variation of $\bar{\omega}_n$ on the boundary $\Gamma_{n-1}^{(\nu)}$ $(\subset \Gamma_{n-1})$ of $F_n^{(\nu)}$. Clearly

$$d_n = d_n^{(1)} + d_n^{(2)} + \cdots + d_n^{(i)}. \tag{3}$$

By the definition of modulus, we have

$$1/\mu_n = 1/\mu_n^{(1)} + 1/\mu_n^{(2)} + \cdots + 1/\mu_n^{(i)}, \tag{4}$$

where $\mu_n = 2\pi/d_n$ and $\mu_n^{(\nu)} = 2\pi/d_n^{(\nu)}$ denote moduli of $F_n - \bar{F}_{n-1}$ and $F_n^{(\nu)}$ respectively. According to (1), the function

$$z_n = \mu_n(\omega_n + i\bar{\omega}_n) + r_{n-1} = x_n + iy_n. \tag{5}$$

[1] SARIO [1] called $e^{\mu}$ the modulus of $G$. This distinction is not essential.

where

$$r_n = \sum_{\nu=1}^{n} \mu_\nu \quad (n \geqq 1), \quad r_0 = 0, \tag{6}$$

maps the open set $F_n - \overline{F}_{n-1}$ with a finite number of suitable slits onto a slit-rectangle $K_n$: $r_{n-1} < x_n < r_n$, $0 < y_n < 2\pi$ in a one to one conformal manner (SARIO [1], p. 11). Accordingly, the function $z = x + iy$ defined by $z_n$ for each $F_n - \overline{F}_{n-1}$ ($n = 1, 2, \ldots$) maps the subsurface $F - \overline{F}_0$ with a finite or an enumerable number of suitable slits onto a union of slit-rectangles: $K = \bigcup_{n=1}^{\infty} K_n$, lying in the domain $0 < x < R = \lim_{n\to\infty} r_n = \sum_{\nu=1}^{\infty} \mu_\nu$, $0 < y < 2\pi$, in a one to one conformal manner. For convenience, we shall call the figure $K$ a *graph* of $F - \overline{F}_0$ by the exhaustion $\{F_n\}$. Similarly we can also define the graph of $F_n - \overline{F}_0$.

**Lemma 1.** *Let $\{F_n\}$ be an exhaustion of a Riemann surface $F$ of parabolic type and $k_\nu$ ($\nu = 1, 2, \ldots$) be an arbitrary sequence of positive numbers. Then there exists a subsequence $\{F_{n_\nu}\}$ ($\nu = 0, 1, \ldots$) which is an exhaustion such that*

$$\mu_{n_\nu} \geqq k_\nu \quad (\nu = 1, 2, \ldots),$$

*where $F_{n_0} = F_0$ and $\mu_{n_\nu}$ denote the modulus of the open set $F_{n_\nu} - \overline{F}_{n_{\nu-1}}$.*

*Proof.* It is obvious that

$$F_n - \overline{F}_j \subset F_n - \overline{F}_0 \quad \text{for } 0 < j < n .$$

Consider two harmonic measures

$$\omega_n^{(j)} = \omega(p, \Gamma_n, F_n - \overline{F}_j) \text{ and } \omega_n^{(0)} = \omega(p, \Gamma_n, F_n - \overline{F}_0) . \tag{7}$$

Then, by the maximum principle, we have

$$\omega_n^{(j)}(p) < \omega_n^{(0)}(p) . \tag{8}$$

Since $F$ is of parabolic type, $\{\omega_n^{(0)}\}$ ($n = 1, 2, \ldots$) converges to a constant zero on $F - \overline{F}_0$. Consequently, for a fixed $j$, $\omega_n^{(j)} \to 0$ as $n \to \infty$. Denote by $\overline{\omega}_n^{(j)}$ the conjugate harmonic function of $\omega_n^{(j)}$ and put

$$\int_{\Gamma_j} d\overline{\omega}_n^{(j)} = d_n^{(j)} > 0 . \tag{9}$$

Then it is easily seen that the modulus $\mu_1^{(j)} = 2\pi / d_n^{(j)}$ of the open set $F_n - \overline{F}_j$ tends to infinity as $n \to \infty$. Accordingly, for any positive number $k$, we can find a number $n$ such that $\mu_n^{(j)} \geqq k$. By repeating the same argument, our assertion is proved.

As an application of the graph of $F_n - \overline{F}_0$ by an exhaustion $\{F_n\}$, we can state

**Theorem 1.** *Let $\mu_n^*$ and $\mu_n$ be the moduli of $F_n - \overline{F}_0$ and $F_n - \overline{F}_{n-1}$ respectively. Then, it holds*

$$\mu_n^* \geqq \mu_1 + \mu_2 + \cdots + \mu_n . \tag{10}$$

*Proof.* Consider the function $w = \omega_n^{(0)} + i\bar{\omega}_n^{(0)}$ (cf. (7)) as a function of $z = x + iy$ in the graph of $F_n - \overline{F}_0$. Then, it is clear that

$$d_n^{(0)} = \int_{x=\lambda} d\bar{\omega}_n^{(0)} \leqq \int_{x=\lambda} |dw| \quad (0 < \lambda < r_n = \sum_{\nu=1}^{n} \mu_\nu) .$$

Schwarz's inequality yields

$$[d_n^{(0)}]^2 \leqq \Big(\int_{x=\lambda} \Big|\frac{dw}{dz}\Big|^2 dy\Big) \cdot \Big(\int_{x=\lambda} dy\Big) = 2\pi \int_{x=\lambda} \Big|\frac{dw}{dz}\Big|^2 dy ,$$

whence, by integration,

$$r_n [d_n^{(0)}]^2 \leqq 2\pi \int_0^{r_n} \int_{x=\lambda} \Big|\frac{dw}{dz}\Big|^2 dy\, d\lambda = 2\pi \int_{\Gamma_0} d\bar{\omega}_n^{(0)} = 2\pi\, d_n^{(0)} .$$

Therefore

$$r_n \leqq 2\pi / d_n^{(0)} = \mu_n^* .$$

Combining the preceding lemma with Sario's theorem [5] which is easily deduced from (10) by Nevanlinna's theorem, we can state

**Theorem 2.** *In order that an open Riemann surface $F$ is of parabolic type, it is necessary and sufficient that there exists an exhaustion $\{F_n\}$ such that $\sum_{n=1}^{\infty} \mu_n = \infty$ where $\mu_n$ denotes the modulus of the open set $F_n - \overline{F}_{n-1}$* (Noshiro [9], Kuroda [2], Sario [9])[1].

**4. Gross' property.** Let $F$ be an open Riemann surface of parabolic type; i. e., $F \in O_G$. Then, by Theorem 2, we can select an exhaustion $\{F_n\}$ of $F$ such that $\sum_{n=1}^{\infty} \mu_n = \infty$, $\mu_n$ denoting the modulus of $F_n - \overline{F}_{n-1}$. Suppose that $w = f(p)$ is non-constant, single-valued and meromorphic on the surface $F$. Then, the space formed by the elements $q = [p, f(p)]$, where $p$ varies on $F$, defines a conformally equivalent covering surface $\Phi$ of the $w$-plane. Clearly the mapping $p \to q$, where $q = [p, f(p)]$, is topological and conformal.

We give a proof[2] for Yûjôbô's theorem[3] which is an extension of a theorem of Tsuji [3].

**Theorem 3.** *The covering surface $\Phi$ belonging to $O_G$ has* Gross' *property.*

*Proof.* Let $q_0 = [p_0, f(p_0)]$ be an arbitrary point on $\Phi$ with projection $w_0 = f(p_0)$. Consider the star-region $H$ formed by the segments from $q_0$ to singular points (algebraic branch points or accessible boundary points of $\Phi$) along all rays: $\arg(w - w_0) = \varphi$ $(0 \leqq \varphi < 2\pi)$ on $\Phi$. We shall show

[1] Sario [3] remarked that a graph $K$ of finite length can be constructed by a suitable choice of an exhaustion of $F$, in the case when $F$ is simply connected and of parabolic type.

[2] Noshiro [9], p. 76.

[3] Yûjôbô [2], Tsuji [13].

that the linear measure of the set $E$ of arguments $\varphi$ of singular rays (by which we understand rays meeting singular points in finite distances) is equal to zero.

Denote by $H_\varrho$ the part of $H$ above a circular disc $|w - w_0| < \varrho$ and by $\Delta_\varrho$ the image of $H_\varrho$ by the mapping $p \to q$. Then $\Delta_\varrho$ is a simply connected domain on the surface $F$. We select as $F_0$ the image of a small circular disc with center $q_0$. Now, we shall use the graph $K$, lying in the half-strip: $0 < x < \infty$, $0 < y < 2\pi$, of the non-compact domain $F - \overline{F}_0$ by the exhaustion $\{F_n\}$ with $\sum_{n=1}^{\infty} \mu_n = \infty$. In the graph $K$ we consider the image $\tilde{\Delta}_\varrho$ of $\Delta_\varrho - \overline{F}_0$ by the function $z(p) = x(p) + iy(p)$, defined in Paragraph 3, and the composed function $w = w(z) = f(p(z))$ defined on $\tilde{\Delta}_\varrho$. Let $\tilde{\Theta}_\lambda$ be the image of the intersection $\Theta_\lambda$ of the niveau curve $C_\lambda$: $x(p) = \lambda$ $(0 < \lambda < \infty)$[1] with $\Delta_\varrho$ by the function $z(p) = x(p) + iy(p)$. We denote by $\theta(\lambda)$ the total length of $\tilde{\Theta}_\lambda$ and by $L(\lambda)$ that of the image of $\tilde{\Theta}_\lambda$ by $w = w(z)$. Then we can apply the method in proving a well-known theorem of GROSS. It is clear that

$$L(\lambda) = \int_{\tilde{\Theta}_\lambda} |w'(z)|\, dy .$$

By Schwarz's inequality

$$[L(\lambda)]^2 \leqq \int_{\tilde{\Theta}_\lambda} |w'(z)|^2 dy \cdot \int_{\tilde{\Theta}_\lambda} dy = \theta(\lambda) \int_{\tilde{\Theta}_\lambda} |w'(z)|^2 dy = \theta(\lambda) \frac{dA(\lambda)}{d\lambda} ,$$

where

$$A(\lambda) = \int_0^\lambda \int_{\tilde{\Theta}\lambda} |w'(z)|^2 dx\, dy .$$

Hence

$$\int_{\lambda_0}^{\lambda} \frac{[L(\lambda)]^2}{\theta(\lambda)} d\lambda \leqq A(\lambda) - A(\lambda_0) \leqq \pi \varrho^2 .$$

Since $\theta(\lambda) \leqq 2\pi$,

$$\int_{\lambda_0}^{\lambda} [L(\lambda)]^2 d\lambda \leqq 2\pi^2 \varrho^2 ,$$

whence follows $\varliminf_{\lambda \to \infty} L(\lambda) = 0$. Accordingly, our assertion is proved by a standard method.

Remark. It is well-known that Iversen's property is a direct result from Gross' property. Thus, Theorem 3 contains a theorem due to STOÏLOW [5]. It is important to notice that we can apply Ahlfors' theory [2] to the covering surface $\Phi$ belonging to $O_G$ with an aid of the preceding lemma, instead of Evans' theorem stated in § 1, II. In particular, *the covering surface $\Phi$ is regularly exhaustible in the sense of* AHLFORS. Furthermore, *if $\Phi$ has an accessible boundary point $\Omega$ with projection $w_0$*

[1] Evidently the niveau curve $C_\lambda$ coincides with $\Gamma_n$ when $\lambda = r_n$ $(n = 0, 1, \ldots)$.

*and if $\Phi_\varrho$ is the $\varrho$-neighborhood of $\Omega$ which is a covering surface of the disc (c): $|w - w_0| < \varrho$, then $\Phi_\varrho$ covers every point infinitely often inside (c) with one possible exception, under the condition that $\Phi_\varrho$ is simply connected* [see Noshiro [9]).

## § 2. Iversen's property of covering surfaces

**1. Iversen's property.** Let $F$ be an open Riemann surface. Let $w = f(p)$ be a non-constant single-valued meromorphic function on $F$. We denote by $\Phi$ the covering surface of the $w$-plane generated by the function $w = f(p)$ on $F$. Let $G$ be any circular disc $|w - w_0| < \varrho$[1] and $\Delta$ be any connected component of the inverse image $f^{-1}(G)$ on $F$. If the set $G - f(\Delta)$ does not contain any (non-degenerate) continuum, we may say that $w = f(p)$ (or the covering surface $\Phi$) has Iversen's *property* (Stoïlow [11]).

Iversen's property can be also explained in various ways. Suppose that there exists a connected component $\Delta$ of the inverse image $f^{-1}(G)$ and that the center $w_0$ of the disc $G$ belongs to the closure $\overline{f(\Delta)}$ of the set $f(\Delta)$, provided that $f(p) \neq w_0$ in $\Delta$. Then, by a standard technic, we can readily prove that there exists an asymptotic path $\mathscr{L}$ in $\Delta$ tending to the ideal boundary of $\Delta$ such that $w_0$ is an asymptotic value of $w = f(p)$ along $\mathscr{L}$; accordingly $w = f(p)$ has Iversen's property. It is obvious that the converse is true.

Stoïlow [1, 8—11] has studied systematically Iversen's property of analytic functions and covering surfaces. One of the most important results of Stoïlow is so-called Stoïlow's *principle on* Iversen's *property* which is described as follows: *If $w = f(p)$ has* Iversen's *property, then either the covering surface $\Phi$ covers every point $w$ the same finite number of times except perhaps a totally disconnected closed set, which consists of the asymptotic values of $w = f(p)$ at the ideal boundary $\Gamma$ of $F$: or the cluster set $C(f)$ (in the large) at $\Gamma$ is total.* We note that the notion of Iversen's property can be easily extended to the case when $w = f(p)$ is a conformal mapping of $F$ into another Riemann surface $W$ (closed or open) (cf. Stoïlow [8, 11], Heins [8, 9]).

**2. Classification of open Riemann surfaces.** Recently many important contributions on the classification of open Riemann surfaces have been made by Sario [2], P. J. Myrberg [4], Ahlfors [5, 6], Ahlfors-Beurling [1], Virtanen [1], Parreau [1], Royden [1, 2, 4], Mori [2—7] and Tôki [2, 3]. For simplicity, we use the following abbreviations.

$H$: non-constant single-valued harmonic,
$A$: non-constant single-valued analytic,
$P$: positive,
$B$: bounded,
$D$: with a finite Dirichlet integral (i. e., Dirichlet-bounded).

[1] In case $w_0 = \infty$, we consider as $G$ a domain $|w| > \varrho$.

We denote by $O_{HB}(O_{HD})$ the class of open Riemann surfaces on which there exists no non-constant bounded (Dirichlet-bounded) harmonic function; and by $O_{HP}, O_{HBD}, O_{AB}$ and $O_{AD}$ the corresponding classes respectively. We have

$$\begin{array}{ccccccc} O_G \subset O_{HP} \subset & O_{HB} & \subset & O_{HD} & = O_{HBD} \\ & \cap & & \cap & \\ & O_{AB} & \subset & O_{AD} & \end{array}$$

For surfaces of finite genus $O_G = O_{HD}$, but TÔKI [3] has proved that for arbitrary Riemann surfaces the inclusions are proper.

**3.** It seems very interesting to investigate interrelations among Iversen's property and the classification of Riemann surfaces. As we have mentioned in § 1, covering surfaces of the $w$-plane belonging to $O_G$ have Gross' property. However, KURAMOCHI [1] pointed out that covering surfaces of the $w$-plane belonging to $O_{HP} - O_G$ have not necessarily Gross' property. Recently, Iversen's property of covering surfaces has been studied by MORI [2, 3], KURODA [7, 8], HEINS [8, 9] and CONSTANTINESCU-CORNEA [1] from the view-point just stated above.

**4.** Let $F$ be an open Riemann surface and $\Delta$ be a non-compact or compact domain on $F$ whose relative boundary $\beta$ with respect to $F$ consists of at most an enumerable number of analytic curves (compact or non-compact) clustering nowhere in $F$. For the sake of simplicity, we call such a domain $\Delta$ a *subregion* on $F$.

Suppose that any non-compact subregion $\Delta$ satisfies the following Lindelöf's maximum principle: If, for any single-valued bounded analytic function $w = \varphi(p)$ on $\Delta \cup \beta$, $\varphi(p)$ is of constant modulus $M$ on $\beta$, then we have $|\varphi(p)| \leqq M$ throughout in $\Delta$. Then, $F$ is said to have *(L)-property.*

**Theorem 1.** *Let $F$ be an open Riemann surface and $w = f(p)$ be a non-constant single-valued meromorphic function on $F$. Then, $w = f(p)$ has* Iversen'*s property, provided that $F$ has (L)-property*[1].

*Proof.* Let $\mathfrak{H}$ be the set of values of $w = f(p)$ on $F$, and $G$ be any circular disc $|w - w_0| < \varrho$ such that the circumference $|w - w_0| = \varrho$ intersects the domain $\mathfrak{H}$. Let $\Delta$ be any connected component, with relative boundary $\beta$, of the inverse image $f^{-1}(G)$. Assume that $f(p) \neq w_0$ in $\Delta$ and $w_0 \notin \overline{f(\Delta)}$. Then, the function $\varphi(p) = (f(p) - w_0)^{-1}$ is single-valued bounded analytic function on $\Delta \cup \beta$ and $|\varphi(p)| = 1/\varrho$ on $\beta$. From $(L)$-property of $F$, it follows that $|\varphi(p)| \leqq 1/\varrho$, i. e. $|f(p) - w_0| \geqq \varrho$ in $\Delta$; this is a contradiction. Thus, we see that the closure of $\mathfrak{H}$ is total and that for any disc $G$: $|w - w_0| < \varrho$, $w_0$ belongs to $\overline{f(\Delta)}$, provided that $f(p) \neq w_0$ in $\Delta$.

Recently, KURODA [7] has introduced a class $O^0_{AB}$ of Riemann surfaces[2]. Let $F$ be an open Riemann surface and $\Delta$ be any subregion of $F$.

[1] Cf. NOSHIRO [4] for some related results.

[2] The importance of such a class was already pointed out by SARIO [2].

If there exists no non-constant single-valued bounded analytic function on $\Delta \cup \beta$, whose real part vanishes on the relative boundary $\beta$ of $\Delta$, then we say $F$ to belong to $O^0_{AB}$.

**Theorem 2.** *In order that F belongs to $O^0_{AB}$, it is necessary and sufficient that F has (L)-property*[1].

*Proof.* Necessity. Suppose that $F$ has no $(L)$-property. Then, there exists some non-compact subregion $\Delta$ with relative boundary $\beta$ and a non-constant single-valued bounded analytic function $w = \varphi(p)$ on $\Delta \cup \beta$ such that $\varphi(p)$ has a constant modulus, say $M$, on $\beta$, but $|\varphi(p_1)| > M$ for some point $p_1 \in \Delta$. Let $\sup_{\Delta} |\varphi(p)| = N$. Consider the component $\Delta_1$ of the inverse image of the annulus $M < |w| < N$ such that $p_1 \in \Delta_1$. Then $\Delta_1$ is non-compact; $|\varphi(p)| = M$ on $\beta_1$, where $\beta_1$ denotes the relative boundary of $\Delta_1$; and $M < |\varphi(p)| < N$ in $\Delta_1$. By using an elementary function $W = \psi(w)$, we can map $M < |w| < N$ one-to-one and conformally onto the interior of an ellipse with a slit on the imaginary axis so as to make the circle $|w| = M$ correspond to the slit. Then, the composed function $W = \psi[\varphi(p)]$ is a non-constant, single-valued bounded analytic function on $\Delta_1 \cup \beta_1$ such that $\Re\, \psi[\varphi(p)] = 0$ on $\beta_1$. Thus, $F$ does not belong to $O^0_{AB}$.

Sufficiency. Suppose that $F$ does not belong to $O^0_{AB}$. Then, there exists a non-compact subregion $\Delta$ and a non-constant single-valued bounded analytic function $w = \varphi(p)$ such that $\Re\, \varphi(p) = 0$ on $\beta$, where $\beta$ denotes the relative boundary of $\Delta$. Without loss of generality, we may assume that $\Re \varphi(p) > 0$ in $\Delta$[2]. Consider the function $w = e^{\varphi(p)}$ on $\Delta \cup \beta$. Then, $|e^{\varphi(p)}| = 1$ on $\beta$ and $|e^{\varphi(p)}| > 1$ in $\Delta$. Accordingly, $F$ has no $(L)$-property.

As an immediate consequence of Theorem 2, we have

**Theorem 3.** *Let F be an open Riemann surface belonging to $O^0_{AB}$ and $w = f(p)$ be a non-constant single-valued meromorphic function on F. Then $w = f(p)$ has* Iversen's *property* (Kuroda [7]).

Let $w = f(p)$ be a non-constant single-valued meromorphic function on $F \in O^0_{AB}$. We now study the set of values of $w = f(p)$, by using Stoïlow's principle. Let $G$ be any circular disc $|w - w_0| < \varrho$ on the $w$-plane and $\Delta$ be any connected component of the inverse image $f^{-1}(G)$. We denote by $\Phi_\Delta$ the covering surface, of the disc $G$, corresponding to $\Delta$. Let $n(w)$ be the number of sheets of $\Phi_\Delta$ above $w \in G$ and $N_\varrho = \sup_G n(w) \leqq +\infty$. We denote by $E_\varrho$ the set $\{w | w \in G, n(w) < N_\varrho\}$.

We prove

**Theorem 4.** *Any compact subset of $E_\varrho$ is a Painlevé null-set* (Kuroda [7]).

---

[1] For the sufficiency, the writer owes to Kuroda.

[2] If $\Re\, \varphi(p_1) > 0$ for some point $p_1$ in $\Delta$, we have only to consider a connected component $\Delta_1$ of the inverse image of the half plane $\Re w > 0$ instead of $\Delta$.

*Proof.* Let $E_n (0 \leqq n < N_\varrho)$ be the set $\{w | w \in G, n(w) \leqq n\}$. Then $E_n$ is relatively closed in $G$, and the sequence $\{E_n\}$ $(0 \leqq n < N_\varrho)$ is monotone non-decreasing. Evidently $E_\varrho = \bigcup_{0 \leqq n < N_\varrho} E_n$. Let $e$ be any compact subset of $E_\varrho$. We shall show that $e$ is a Painlevé null-set. Contrary to the assertion, we assume that $e$ is not a null-set. Then, there exists an integer $n (\geqq 0)$ such that the compact set $e_n = e \cap E_n$ is not a null-set. Let us denote by $m$ the smallest of all such non-negative integers $n$. Then, $e_{m-1}$ is a null-set, but $e_m$ is not. By Lindelöf's covering theorem, we see that there exists a point $w_1 \in (e_m - e_{m-1})$ such that for any positive number $\varrho_1$ the intersection of $e_m$ with the closed disc $|w - w_1| \leqq \varrho_1$ is not a null-set, since $e_{m-1}$ is a null-set. By a localization of Stoïlow's principle, $w_1$ is a cluster value of $w = f(p)$ at the ideal boundary of $\Delta$. Consequently, if we select a positive number $\sigma_1$ suitably small, then the inverse image $f^{-1}(G_1)$ of $G_1$: $|w - w_1| < \sigma_1$ in $\Delta$ consists of at most $m$ relatively compact components (islands) and at least one relatively non-compact component (peninsula) $\Delta_1$. Since the set of values of $w = f(p)$ in $\Delta_1$ does not cover any point of a totally disconnected compact set $\mathscr{E} = e_m \cap (w| \, |w - w_1| \leqq \sigma_2; \sigma_2 < \sigma_1)$, which is not a null-set, we arrive at a contradiction, as $F$ belongs to $O^0_{AB}$[1].

Remark. It is easy to prove that $O^0_{AB} \subset O_{AB}$ (Kuroda [7], p. 38). On the other hand, P. J. Myrberg [4] has given a very important example of a covering Riemann surface $\Phi$ of infinite genus which belongs to $O_{AB}$ and has not Iversen's property. From this fact and Theorem 3, it follows that the inclusion $O^0_{AB} \subset O_{AB}$ is proper. There is an open problem *to settle whether a Riemann surface of planar character belonging to* $O_{AB}$ *has* Iversen's *property*. Kuroda [7] has also proved that $O_{HB} \subset O^0_{AB}$ and the inclusion is proper. However, this is closely related to results of Mori [2, 3], Kuramochi [3] and others. We shall discuss these interrelations in the following § 3.

## § 3. Boundary theorems on open Riemann surfaces

1. Let $F$ be an open Riemann surface and $\Delta$ be a subregion of $F$ with relative boundary $\beta$. If there exists no non-constant bounded (Dirichlet-bounded) harmonic function on $\Delta \cup \beta$ which vanishes continuously on $\beta$, then we say that $\Delta$ belongs to $SO_{HB} (SO_{HD})$. Mori [3] proved that

$$SO_{HB} \subset SO_{HD} = SO_{HBD}$$

and the inclusion is proper. It is easy to show that $\Delta \in SO_{HB}$ if and only if the ideal boundary $\gamma$ of $\Delta$ is of harmonic measure zero relative to $\Delta$.

Kuramochi [3] has obtained

[1] Since $\mathscr{E}$ is not a null-set, there exists a non-constant single-valued bounded analytic function $\psi(w)$ in $G_1 - \mathscr{E}$ such that $\Re\psi(w)$ vanishes continuously on $|w - w_1| = \sigma_1$. Considering the composed function $\psi[f(p)]$ inside $\Delta_1$, we conclude that $\psi[f(p)]$ is identically zero, as $F$ belongs to $O^0_{AB}$.

**Theorem 1.** *Let $F$ be an open Riemann surface belonging to $O_{HB}(O_{HD})$. Let $\Delta$ be a non-compact subregion of $F$ with compact relative boundary $\beta$ such that $\Delta \notin SO_{HB} = SO_{HD}$*[1]. *Then, $\Delta$ belongs to $O_{AB}(O_{AD})$.*

CORNEA [1] has given a simple proof for this theorem, with the aid of a theorem of SARIO [6, 10]. However, if we make use of a theorem of NEVANLINNA [5], then Cornea's method is also applicable to the case of a subregion $\Delta$ with non-compact relative boundary $\beta$ as far as we are concerned with bounded harmonic functions.

**Theorem 2.** *Let $F$ be an open Riemann surface belonging to $O_{HB}$. Let $\Delta$ be a non-compact subregion of $F$ with relative boundary $\beta$ such that $\Delta \notin SO_{HB}$. Then $\Delta$ belongs to $O_{AB}$*[2].

*Proof.* Let $w = f(p) = u(p) + iv(p)$ be a single-valued bounded analytic function in $\Delta$. Without loss of generality, we may suppose that $w = f(p)$ is also analytic on the relative boundary $\beta$[3]. Let $\{F_n\}$ be an exhaustion of $F$. We denote by $\Gamma_n$ the boundary of $F_n$. Let $u_n$ be the harmonic function on $F_n \cap \Delta$ such that $u_n = u$ on $\beta \cap F_n$ and $u_n = 0$ on $\Gamma_n \cap \Delta$. Since $u$ is bounded on $\beta$, we can choose a subsequence of $\{u_n\}$ which converges uniformly on any compact subset of $\Delta \cup \beta$. We denote by $u^*$ the limiting function. Obviously $u^*$ is bounded and harmonic on $\Delta \cup \beta$ such that $u^* = u$ on $\beta$. Let us put $u(p_0) - u^*(p_0) = a\omega(p_0)$ for some point $p_0$ in $\Delta$, where $\omega = \omega(p, \Delta)$ denotes the harmonic measure of the ideal boundary $\gamma$ of $\Delta$ relative to $\Delta$. By a theorem of NEVANLINNA [5], we have

$$u - u^* - a\omega \equiv 0 \text{ on } \Delta \cup \beta\,^4.$$

Accordingly

$$u - a \equiv u^* - a(1 - \omega);$$

hence

$$|u - a| \leqq M(1 - \omega) \text{ in } \Delta, \tag{1}$$

provided that $|u - a| \leqq M$ on $\beta$.

Now, we consider the universal covering surface $\tilde{\Delta}$ of $\Delta$. We map $\tilde{\Delta}$ by $p = \varphi(\zeta)$ conformally onto the unit disc $D$: $|\zeta| < 1$. Let $E_0$ be the image of $\beta$ and $E_1$ be its complement with respect to $|\zeta| = 1$. We note that $\omega(\varphi(\zeta), \Delta)$ coincides with the harmonic measure $\omega(\zeta, E_1, D)$ of $E_1$ with

[1] In the particular case where the relative boundary $\beta$ of $\Delta$ is compact, we should note that $\Delta \in SO_{HB}$ if and only if $\Delta \in SO_{HD}$.

[2] We shall discuss later the case of $F \in O_{HD}$. In fact, Theorem 2 remains valid even if we replace the letter "$B$" by "$D$".

[3] For that purpose, we have only to replace $\Delta$ by a connected component of the open set $\{p \mid \varepsilon < \omega(p, \Delta) < 1\}$, where $\omega(p, \Delta)$ is the harmonic measure of the ideal boundary $\gamma$ of $\Delta$ relative to $\Delta$ and $0 < \varepsilon < 1$.

[4] The identity follows from a well-known theorem: If there exists two subregions $\Delta_1$ and $\Delta_2$ on $F$ such that $\Delta_1 \notin SO_{HB}$, $\Delta_2 \notin SO_{HB}$ and $\Delta_1 \cap \Delta_2 = \emptyset$, then $F \notin O_{HB}$ (NEVANLINNA [5], BADER-PARREAU [1], PARREAU [1], ROYDEN [1], MORI [3]). As another consequence of this theorem, it should be noted that the open set $\{p \mid \varepsilon < \omega(p, \Delta) < 1\}$ consists of a single component.

respect to $D^1$. Consequently, $E_1$ has positive measure and $\omega(\varphi(\zeta), \Delta)$ has angular limit 1 almost everywhere on $E_1$; hence, by (1), $u(\varphi(\zeta))$ has angular limit $a$ almost everywhere on $E_1$. By an entirely similar argument, we see that $v(\varphi(\zeta))$ has the same angular limit, say $b$, almost everywhere on $E_1$. Thus, $f(\varphi(\zeta))$ has angular limit $\alpha = a + ib$ almost everywhere on $E_1$. By Lusin-Privaloff's theorem, $f(\varphi(\zeta))$ must be identically constant.

As an immediate consequence of Theorem 2, we obtain

**Theorem 3.** *Let $F$ be an open Riemann surface belonging to $O_{HB}$ and $w = f(p)$ be a non-constant single-valued meromorphic function $F$. Let $G$ be any circular disc $|w - w_0| < \varrho$ on the $w$-plane and $\Delta$ be any connected component of the inverse image $f^{-1}(G)$. Then, $\Delta$ belongs to $SO_{HB}$* (Mori [2])[2].

Remark. Let $p = \varphi(\zeta)$ be the function considered in the proof of Theorem 2. Then $[f(\varphi(\zeta)) - w_0]/\varrho$ is a function of class $(U)$ in Seidel's sense. Accordingly, we can apply the theory of functions of class $(U)$ to the present case. For example, the covering surface $\Phi$ of the $w$-plane generated by $w = f(p)$ has Iversen's property. Furthermore, corresponding to Theorem 4, § 2, we get the following result: The covering surface $\Phi_\Delta$, corresponding to $\Delta$, of $G$ covers every point of $G$ the same number $N_\varrho$ (finite or infinite) of times except for at most an $F_\sigma$ set[3] of capacity zero.

**Theorem 4.** $O_{HB} \subset O^0_{AB}$[4].

*Proof.* Let $F$ be an open Riemann surface belonging to $O_{HB}$ and $\Delta$ be any non-compact subregion with relative boundary $\beta$. Let $w = f(p)$ be any single-valued bounded analytic function on $\Delta \cup \beta$ whose real part vanishes on $\beta$. If $\Delta$ belongs to $SO_{HB}$, then it is evident that $\Re f(p)$ is identically constant. If $\Delta$ does not belong to $SO_{HB}$, then, by Theorem 2, $\Delta$ must belong to $O_{AB}$; therefore $w = f(p)$ must be identically constant. Thus, we have $O_{HB} \subset O^0_{AB}$.

2. Heins [8, 9] has established an important theory on the Lindelöf's principle and Lindelöfian mappings which generalizes Lehto's theory of meromorphic functions of bounded type (cf. § 5, III) to the case of open Riemann surfaces (see also Parreau [2, 3]). Heins' methods and results are also very powerful for the study of boundary behaviours.

Let $F$ be an open Riemann surface and $\mathfrak{U}$ be the class of non-negative harmonic functions on $F$. A member $u$ of $\mathfrak{U}$ is called *quasi-bounded* provided that $u$ admits a representation as the limit of a monotone non-decreasing sequence of non-negative bounded harmonic functions on $F$. A member $u$ of $\mathfrak{U}$ is called *singular* provided that the only bounded member

[1] Cf. Constantinescu-Cornea [1], p. 194.

[2] In other words, $w = f(p)$ is locally of type-Bl in Heins' sense. We shall discuss later conformal mapping locally of type-Bl.

[3] Here, an $F_\sigma$ set means a union of at most countably many closed sets relative to $G$.

[4] The fact that the covering surface $\Phi$ of the $w$-plane belonging to $O_{HB}$ has Iversen's property follows also from this theorem, if we make use of Theorem 3, § 2.

of $\mathfrak{U}$ dominated by $u$ is 0. A member $u$ ($>0$) of $\mathfrak{U}$ is called *minimal* provided that the members of $\mathfrak{U}$ dominated by $u$ are of the form $cu$ where $c$ is a constant. PARREAU [1] has proved that *each $u$ of $\mathfrak{U}$ is representable uniquely as the sum of a quasi-bounded and a singular member of $\mathfrak{U}$.*

We consider two open Riemann surfaces $F$ and $W$ of hyperbolic type. Let $f$ be a conformal mapping (not necessarily univalent) of $F$ into $W$. Let $\mathcal{G}_F$ and $\mathcal{G}_W$ denote Green's functions of $F$ and $W$ respectively. Then we have

$$\mathcal{G}_W(f(p); q) = \sum_{f(r)=q} n(r)\, \mathcal{G}_F(p; r) + u_q(p), \tag{2}$$

where $n(r)$ denotes the multiplicity of $f$ at $r \in F$ and $u_q(p)$ is *the greatest harmonic minorant* of $\mathcal{G}_W(f(p); q)$ on $F$. HEINS [8] has proved that $u_q(p)$ *is quasi-bounded except for a set of $q$ of capacity zero and* that *the quasi-bounded component of $u_q(p)$ is either positive on $F \times W$ or else identically zero on $F \times W$.* According to HEINS [8], we call $f$ *of type-Bl* if the second alternative holds.

We now consider two Riemann surfaces $F$ and $W$, without any restriction on $F$ and $W$. We call $f$ *of type-Bl at $q \in W$* provided that there exists a simply connected Jordan domain $G$ which satisfies: (i) $q \in G \subset W$, (ii) $f^{-1}(G) \neq \emptyset$, (iii) for each component $\Delta$ of $f^{-1}(G)$, the restriction of $f$ to $\Delta$ is a mapping of type-Bl of $\Delta$ into $G$. We say that $f$ is *locally of type-Bl* provided that $f$ is of type-Bl at every point of $W$.

**Theorem 5.** *Let $F$ be an open Riemann surface and $W$ be an open or closed Riemann surface. Let $f$ be a conformal mapping of $F$ into $W$. Then, in order that $f$ is locally of type-Bl, it is necessary and sufficient that, for every compact subregion $G$ of $W$ (such that $\overline{G} \neq W$ in case $W$ is closed), each component $\Delta$ of $f^{-1}(G)$ belongs to $SO_{HB}$* (MATSUMOTO [1]).

*Proof.* The sufficiency is evident. Suppose that $f$ is locally of type-Bl. We choose a compact subregion $G_1$ of $W$ which contains $\overline{G}$ in its interior. Let $\Delta$ be any component of $f^{-1}(G)$ and $\Delta_1$ be the component of $f^{-1}(G_1)$ containing $\Delta$. Contrary to the assertion, assume that $\Delta \notin SO_{HB}$, and denote by $\omega = \omega(p, \Delta)$ the harmonic measure of the ideal boundary of $\Delta$. We denote by $u$ the standard subharmonic extension of $\omega = \omega(p, \Delta)$ to $\Delta_1$ such that $u = \omega$ in $\Delta$ and $u = 0$ in $\Delta_1 - \Delta$. Put $A = \min_{s \in \overline{G}} \mathcal{G}_{G_1}(s; q)$, where $q$ is some point of $G_1$. Then, we have $A u \leqq \mathcal{G}_{G_1}(f_{\Delta_1}(p), q)$ in $\Delta_1$ where $f_{\Delta_1}$ denotes the restriction of $f$ to $\Delta_1$. Since the least harmonic majorant of $Au$ in $\Delta_1$ is positive and bounded, and dominated by $\mathcal{G}_{G_1}(f_{\Delta_1}(p); q)$, the mapping $f_{\Delta_1}$ of $\Delta_1$ into $G_1$ cannot be of type-Bl; this contradicts a theorem of HEINS ([8], p. 466).

Remark. Taking as $W$ the complex $w$-plane and using this theorem, we can rewrite Theorem 3 as follows: *If $w = f(p)$ is a non-constant single-valued meromorphic function on $F \in O_{HB}$, then $w = f(p)$ is locally of type-Bl.*

**3.** To discuss the interrelation between results of HEINS and those of CONSTANTINESCU and CORNEA [1], we need some preliminaries.

Let $\Delta$ be a subregion on an open Riemann surface $F$ with relative boundary $\beta$ (compact or non-compact). We denote by $\mathfrak{U}$ the class of non-negative harmonic functions on $F$ and by $\mathfrak{U}_\Delta$ the class of non-negative harmonic functions with domain $\Delta$ which vanish continuously on $\beta$. For each $u \in \mathfrak{U}$, we denote by $\lambda_\Delta(u)$ the upper envelope of the set of members of $\mathfrak{U}_\Delta$ which are dominated by $u$ in $\Delta$; it is evident that $\lambda_\Delta(u)$ is itself a member of $\mathfrak{U}_\Delta$. The mapping $\lambda_\Delta$ of $\mathfrak{U}$ into $\mathfrak{U}_\Delta$ is additive and homogeneous. It is easy to show by an example that $\lambda_\Delta$ is not necessarily univalent. Next, we consider for each $U \in \mathfrak{U}_\Delta$ its standard subharmonic extension $\tilde{U}$ such that $\tilde{U} = U$ in $\Delta$ and $\tilde{U} = 0$ in $F - \Delta$. Let $\mathfrak{U}_\Delta^*$ be the subset of $\mathfrak{U}_\Delta$ which consist of those $U$ such that $\tilde{U}$ admits a harmonic majorant on $F$. For each $U \in \mathfrak{U}_\Delta^*$, we denote by $\mu_\Delta(U)$ the least harmonic majorant of $\tilde{U}$ on $F$; obviously $\mu_\Delta(U) \in \mathfrak{U}$. The mapping $\mu_\Delta$ of $\mathfrak{U}_\Delta^*$ into $\mathfrak{U}$ is additive and homogeneous. These two mappings $\lambda_\Delta$ and $\mu_\Delta$ have been introduced by HEINS [8] and others[1]. Concerning the mappings $\lambda_\Delta$ and $\mu_\Delta$, the following properties (a), (b) and (c) have been proved by HEINS (see [8], p. 442).

(a) For $U \in \mathfrak{U}_\Delta^*$, we have

$$\lambda_\Delta(\mu_\Delta(U)) = U ;$$

i. e. the mapping $\mu_\Delta$ is univalent and hence the restriction of $\lambda_\Delta$ to $\mu_\Delta(\mathfrak{U}_\Delta^*)$ is univalent.

(b) If $u \in \mathfrak{U}$ is dominated by some member of $\mu_\Delta(\mathfrak{U}_\Delta^*)$, then $u$ belongs to $\mu_\Delta(\mathfrak{U}_\Delta^*)$.

(c) Let $U$ belong to $\mathfrak{U}_\Delta^*$. Then $\mu_\Delta(U)$ is quasi-bounded (singular, minimal) in $F$ if and only if $U$ is quasi-bounded (singular, minimal) in $\Delta$.

As complements of these properties, we give some lemmas obtained by MATSUMOTO [1].

**Lemma 1.** (Complement of (c)). *Let $u \in \mathfrak{U}$ be minimal. Then, $\lambda_\Delta(u)$ is also minimal provided that $\lambda_\Delta(u)$ is positive in $\Delta$.*

*Proof.* We note that $\lambda_\Delta(u) \in \mathfrak{U}_\Delta^*$, as $\lambda_\Delta(u) \leqq u$ in $\Delta$. Obviously $\mu_\Delta(\lambda_\Delta(u)) \leqq u$ in $F$. From the minimality of $u$, it follows that $\mu_\Delta(\lambda_\Delta(u)) = cu$ $(0 < c \leqq 1)$. Hence $\lambda_\Delta(\mu_\Delta(\lambda_\Delta(u))) = c\lambda_\Delta(u)$; i. e. $\lambda_\Delta(u) = c\lambda_\Delta(u)$. Whence follows that $c = 1$ as $\lambda_\Delta(u) > 0$. Consequently $\mu_\Delta(\lambda_\Delta(u)) = u$. Then, $\lambda_\Delta(u)$ is minimal by (c).

[1] We adopt here Heins' notations. The mappings $\lambda_\Delta$ and $\mu_\Delta$ have been also studied by PARREAU ([1], pp. 44—45), KURAMOCHI [2] and CONSTANTINESCU-CORNEA ([1], pp. 186—200). However, they used different terminologies and notations. CONSTANTINESCU and CORNEA called $\lambda_\Delta$ and $\mu_\Delta$ the Inextremisation $I$ and Extremisation $E$ with respect to the pair $(\Delta, F)$ respectively.

**Lemma 2.** *Let $u$ and $u_n$ $(n = 1, 2, \ldots)$ belong to $\mathfrak{U}$ and $u = \sum_{n=1}^{\infty} u_n$. Then,*

$$\lambda_\Delta(u) = \sum_{n=1}^{\infty} \lambda_\Delta(u_n) . \tag{3}$$

*Similarly, let $U$ and $U_n$ $(n = 1, 2, \ldots)$ belong to $\mathfrak{U}_\Delta^*$ and $U = \sum_{n=1}^{\infty} U_n$. Then*

$$\mu_\Delta(U) = \sum_{n=1}^{\infty} \mu_\Delta(U_n)^1 . \tag{4}$$

*Proof.* For any natural number $m$, we have evidently

$$\lambda_\Delta(u) \geqq \sum_{n=1}^{m} \lambda_\Delta(u_n) ;$$

hence

$$\lambda_\Delta(u) \geqq \sum_{n=1}^{\infty} \lambda_\Delta(u_n) .$$

Let $p$ be any fixed point on $F$ and $\varepsilon$ be any positive number. We choose a natural number $m = m(p, \varepsilon)$ such that $\sum_{n=m+1}^{\infty} u_n(p) < \varepsilon$. Then

$$\lambda_\Delta(u)(p) = \lambda_\Delta\left(\sum_{n=1}^{m} u_n\right)(p) + \lambda_\Delta\left(\sum_{n=m+1}^{\infty} u_n\right)(p)$$
$$< \left(\sum_{n=1}^{m} \lambda_\Delta(u_n)\right)(p) + \varepsilon \leqq \left(\sum_{n=1}^{\infty} \lambda_\Delta(u_n)\right)(p) + \varepsilon ;$$

by making $\varepsilon \to 0$, we have

$$\lambda_\Delta(u)(p) \leqq \left(\sum_{n=1}^{\infty} \lambda_\Delta(u_n)\right)(p) .$$

Thus, it is proved that (3) holds. To prove (4), put $v = \mu_\Delta(U)$ and $v_n = \mu_\Delta(U_n)$. Evidently $\sum_{n=1}^{\infty} v_n \leqq v$. By (b), there exists a member $V$ of $\mathfrak{U}_\Delta^*$ such that $\sum_{n=1}^{\infty} v_n = \mu_\Delta(V)$. Apply the relation (3) to this equality. Then $\sum_{n=1}^{\infty} \lambda_\Delta(v_n) = \lambda_\Delta(\mu_\Delta(V))$; i. e. $\sum_{n=1}^{\infty} U_n = V$ by (a). Hence $U = V$. Thus, it is proved that (4) is valid.

Now, we give a brief account on the concept of an *indivisible set* of points of the ideal boundary of a Riemann surface which has been recently introduced by CONSTANTINESCU and CORNEA [1]. Let $F$ be an open Riemann surface of hyperbolic type and $p = \varphi(\zeta)$ be the univalent conformal mapping of $D$: $|\zeta| < 1$ onto the universal covering surface $\tilde{F}$ of $F$.

[1] In a special case, this lemma was proved by CONSTANTINESCU-CORNEA ([1], p. 189).

Let $p_0$ be an arbitrary point of $F$. Then, we have

$$\mathscr{G}_F(\varphi(\zeta), p_0) = \sum_{\varphi(\zeta_\nu)=p_0} \mathscr{G}_D(\zeta; \zeta_\nu) = \sum_{\nu=1}^{\infty} \log \left| \frac{1-\bar{\zeta}_\nu \zeta}{\zeta - \zeta_\nu} \right| = \log \left| \frac{1}{B(\zeta)} \right|, \quad (5)$$

where $B(\zeta)$ is the Blaschke product (see NEVANLINNA [7], p. 207). Since $B(\zeta)$ is a function of class $(U)$ in Seidel's sense, $\mathscr{G}_F(\varphi(\zeta), p_0)$ has angular limit 0 almost everywhere on $|\zeta| = 1$. Let us denote by $\mathscr{M}$ the set of points $\zeta = e^{i\vartheta}$ for which $\mathscr{G}_F(\varphi(\zeta), p_0)$ has angular limit 0[1]. As is well-known, the function $\varphi(\zeta)$ is automorphic with respect to the Fuchsian group $(T)$ corresponding to the group of covering transformations of $\tilde{F}$. Obviously $\mathscr{M} = T(\mathscr{M})$ for any linear transformation $T \in (T)$. We call two points $\zeta_1$ and $\zeta_2$ belonging to $\mathscr{M}$ equivalent provided that there exists a $T \in (T)$ such that $\zeta_2 = T(\zeta_1)$; we can decompose the set $\mathscr{M}$ into equivalence classes. We say that an equivalence class defines a point $\mathfrak{p}$ on the ideal boundary $\Gamma$ of $F$. The set $E_{\mathfrak{p}}$ of points $\zeta = e^{i\vartheta}$ which constitute the equivalence class is called the image of $\mathfrak{p}$. A set $\gamma$ of points on $\Gamma$ is a certain set of equivalence classes and the set $E_\gamma$ of all the points $\zeta = e^{i\vartheta}$ contained in these equivalence classes is called the image of $\gamma$ on $|\zeta| = 1$. A set $\gamma$ of points on $\Gamma$ is said to be *measurable* provided that its image $E_\gamma$ on $|\zeta| = 1$ is measurable. By definition, $\omega(p, \gamma, F) = \omega(\varphi^{-1}(p), E_\gamma, D)$. We call a set $\gamma$ on $\Gamma$ *indivisible* provided that $E_\gamma$ is measurable, $\omega(p, \gamma, F) > 0$ and $\gamma$ cannot be decomposed into two disjoint measurable sets $\gamma_1, \gamma_2$ of positive harmonic measure (CONSTANTINESCU-CORNEA [1], p. 178).

**Theorem 6.** *Let $\gamma$ be an indivisible set of points on the ideal boundary $\Gamma$ of $F$ and $E_\gamma$ its image. If $u$ is harmonic on $F$ such that $u(\varphi(\zeta))$ has angular limits $u(\varphi(e^{i\vartheta}))$ almost everywhere on $|\zeta| = 1$, then the angular limit function $u(\varphi(e^{i\vartheta}))$ is constant almost everywhere on $E_\gamma$* (CONSTANTINESCU-CORNEA [1][2]).

Furthermore, CONSTANTINESCU and CORNEA ([1], p. 182) have proved: *Every bounded minimal harmonic function $u$ corresponds to an indivisible set and its converse is valid.*

On the other hand, HEINS [9] has introduced a class $O_L$ of Riemann surfaces on which there exists no non-constant single-valued Lindelöfian meromorphic function[3] and proved the proper inclusion $O_{HB} \subset O_L \subset O_{AB}$.

We now prove

[1] We can prove that the set $\mathscr{M}$ does not depend on the choice of $p_0$.

[2] For the proof, see CONSTANTINESCU-CORNEA [1], p. 181, Satz 2.

[3] We say that a conformal mapping of a Riemann surface $F$ into another Riemann surface $W$ is Lindelöfian provided that

$$\sum_{f(r)=q} n(r)\, \mathscr{G}_F(p, r) < \infty$$

for $f(p) \neq q$.

**Theorem 7.** *Let $F$ be a Riemann surface belonging to $O_{HB}$ and $\Delta$ be a subregion on $F$ which does not belong to $SO_{HB}$. Then $\Delta$ belongs to $O_L$*[1] (KURAMOCHI [4]).

*Proof.* Since $u \equiv 1$ is minimal in $F$ and $\lambda_\Delta(1) \equiv \omega(p, \Delta)$, which is no other than the harmonic measure of the ideal boundary $\gamma$[2] of $\Delta$ (cf. the proof of Theorem 2), is positive, $\omega(p, \Delta)$ is minimal in $\Delta$ by Lemma 1. Let $p = \varphi(\zeta)$ be the univalent conformal mapping of $D$: $|\zeta| < 1$ onto the universal covering surface $\tilde{\Delta}$ of $\Delta$. Let $E_1$ be the set of points $\zeta = e^{i\vartheta}$ for which $\omega(\varphi(\zeta), \Delta)$ has angular limit 1. Then, $E_1$ is the image of the indivisible set $\gamma$. Consider any single-valued Lindelöfian meromorphic function $w = f(p)$ in $\Delta$. Then, the composed function $w = f(\varphi(\zeta))$ is also Lindelöfian in $|\zeta| < 1$ by Heins' composition theorem ([9], Theorem 11. 1, p. 440); hence $w = f(\varphi(\zeta))$ is a meromorphic function of bounded type in $|\zeta| < 1$. By Fatou-Nevanlinna's theorem, $w = f(\varphi(\zeta))$ has angular limits $f(\varphi(e^{i\vartheta}))$ almost everywhere on $|\zeta| = 1$. By Theorem 6, the angular limit function $f(\varphi(e^{i\vartheta}))$ is constant almost everywhere on $E_1$ of positive measure. Consequently, $f(\varphi(\zeta))$ is identically constant in $|\zeta| < 1$ by Lusin-Privaloff's theorem.

As an immediate consequence of Theorem 7, we have

**Theorem 8.** $O_{HB} \subset O_L$.[3]

Let us now consider the set $\Gamma(\mathcal{M})$ of all the points of the ideal boundary $\Gamma$ of an open Riemann surface $F$ of hyperbolic type; obviously, the set $\mathcal{M}$ (defined before) on $|\zeta| = 1$ is the image of $\Gamma(\mathcal{M})$. It is readily shown that the set $\Gamma(\mathcal{M})$ contains at most a countable number of indivisible sets. We call that $F$ belongs to $O_{HB_n}$ $(1 \leqq n \leqq \infty)$ provided that $\Gamma(\mathcal{M})$ consists of at most $n$ indivisible sets $\gamma_k$ (disregarding a set of harmonic measure zero) (CONSTANTINESCU-CORNEA [1], p. 222). Evidently

$$O_{HB} = O_{HB_1}\,; \tag{6}$$

$$O_{HB_1} \subset O_{HB_2} \subset \cdots \subset O_{HB_n} \subset \cdots \subset O_{HB_\infty}.^{4} \tag{7}$$

Applying Lemma 2, we can generalize Theorem 7 as follows:

**Theorem 9.** *Suppose that $F \in O_{HB_n}$ $(1 \leqq n \leqq \infty)$ and $\Delta$ is a subregion on $F$ which does not belong to $SO_{HB}$. Then $\Delta \in O_L$.*

---

[1] Clearly this is an extension of Theorem 2. The following proof is due to MATSUMOTO [2].

[2] Note that the boundary of $\Delta$ consists of the relative boundary $\beta$ and the ideal boundary $\gamma$.

[3] For the proof of the inclusion, we have only to note that $\mathcal{G}_\Delta(p, r) < \mathcal{G}_F(p, r)$ provided that $F \in O_{HB} - O_G$ and $\Delta$ be the remaining subregion obtained by excluding a Jordan closed domain from $F$.

[4] The equality (6) follows from a theorem of NEVANLINNA [5]. It can be shown by examples that the inclusions (7) are proper (see CONSTANTINESCU-CORNEA [1], p. 230).

*Proof.* Suppose that $\Gamma(\mathcal{M})$ consists of just $m$ $(\leq n)$ indivisible sets $\gamma_k$ $(k=1,2,\ldots,m)$. We denote by $\omega_k \equiv \omega(p, \gamma_k, F)$ the harmonic measure of $\gamma_k$ with respect to $F$. Then, each $\omega_k$ is minimal and $\sum_{k=1}^{m} \omega_k \equiv 1$. Since $\Delta$ does not belong to $SO_{HB}$, $\lambda_\Delta(1) \equiv \omega(p, \Delta) = \sum_{k=1}^{m} \lambda_\Delta(\omega_k)$ is positive and, for some $k$, $\lambda_\Delta(\omega_k)$ is positive and minimal on $\Delta$ by Lemma 2. Let $E_1$ be the set of the points $\zeta = e^{i\vartheta}$ for which $\lambda_\Delta(\omega_k(\varphi(\zeta)))$ has angular limit 1. Repeating an entirely similar argument used in the proof of Theorem 7, we have our assertion.

Remark. It follows that under the same hypothesis of Theorem 9 $\Delta$ belongs to $O^0_{AB}$. Consequently, we get the following proper inclusions:

$$O_{HB} = O_{HB_1} \subset O_{HB_2} \subset \cdots \subset O_{HB_n} \subset \cdots \subset O_{HB_\infty} \subset (O^0_{AB} \cap O_L)\,.^1$$

**4.** We shall give a brief discussion on the boundary properties of Riemann surfaces belonging to $O_{HD}$ or some related classes. First of all, an example of Tôki [2] shows that *there exits a covering Riemann surface belonging to $O_{HD}$ which has no Iversen's property*. Let $F$ be an open Riemann surface of hyperbolic type. Consider the class $HD$ of Dirichlet-bounded harmonic functions on $F$. Let $\gamma$ be a set of points on the ideal boundary $\Gamma$ of $F$ in the sense of Constantinescu and Cornea and $E_\gamma$ be its image on $|\zeta| = 1$. We call $\gamma$ *HD-indivisible* provided that $\gamma$ is of positive harmonic measure and, for each member $u \in HD$, the angular limit function $u(\varphi(e^{i\vartheta}))$ is constant almost everywhere on $E_\gamma$ (Constantinescu-Cornea [1], p. 200). Such a $HD$-indivisible set $\gamma$ is called *maximal* provided that there exists no $HD$-indivisible set $\bar{\gamma}$ such that $\gamma \subset \bar{\gamma}$ and $\omega(p, \bar{\gamma} - \gamma, F) > 0$.

Let us consider the subclass $\mathfrak{V}$ of $\mathfrak{U}$, each member $u$ of which admits a representation as the limit of a monotone non-increasing sequence of non-negative Dirichlet-bounded harmonic functions on $F$. We say that a positive harmonic function $u \in \mathfrak{V}$ is *HD-minimal* on $F$ provided that the members of $\mathfrak{V}$ dominated by $u$ are of the form $cu$ where $c$ is a constant. Let $\Delta$ be a subregion on $F$. In an entirely similar manner, we can define the subclass $\mathfrak{V}_\Delta$ of $\mathfrak{U}_\Delta$ and the $HD$-minimality of a member of $\mathfrak{V}_\Delta$. Constantinescu and Cornea ([1], p. 209) have proved that if $u$ and $v$ belong to $\mathfrak{V}(\mathfrak{V}_\Delta)$, then the greatest harmonic minorant $u \wedge v$ of the superharmonic function $\min(u, v)$ and the least harmonic majorant $u \vee v$ of the subharmonic function $\max(u, v)$ also belong to $\mathfrak{V}(\mathfrak{V}_\Delta)$.

---

[1] Kuroda ([7], p. 48) has remarked that there exists a plane domain belonging to $O^0_{AB}$ whose boundary is of positive capacity; hence $O^0_{AB} \subset O_L$. On the other hand, as is easily seen from Theorem 7, there exists a covering surface belonging to $O_L$ which has no Iversen's property; hence $O_L \subset O^0_{AB}$.

**Lemma 3.** *Let $u$ be an $\underline{HD}$-minimal function on $F$ and let $\Delta$ be a subregion not belonging to $SO_{HD}$. Then $\lambda_\Delta(u)$ is also $\underline{HD}$-minimal on $\Delta$ provided that there exists a positive Dirichlet-bounded harmonic function $U$ belonging to $\mathfrak{U}_\Delta^*$ such that $\mu_\Delta(U)$ dominates $u$ on $F$* (MATSUMOTO [2]).

*Proof.* By the property (b), there exists a positive harmonic function $V \in \mathfrak{U}_\Delta^*$ such that $\mu_\Delta(V) = u$, since $\mu_\Delta(U) \geqq u$. Hence $U \geqq V$ and $u \geqq u \wedge U \geqq V$ on $\Delta$. By the definition of the mapping $\lambda_\Delta$, we see that $u \wedge U = V$ on $\Delta$, since $u \wedge U$ vanishes continuously on $\beta$ and $\lambda_\Delta(u) = V$. Thus, $V$ belongs to $\mathfrak{V}_\Delta$. Let $V'$ be a member of $\mathfrak{V}_\Delta$ dominated by $V$. Suppose $V'$ admits a representation as the limit of a monotone non-increasing sequence $\{V_k'\}$ of non-negative Dirichlet-bounded harmonic functions on $\Delta$. Then, the sequence $\{U \wedge V_k'\}$ converges also to the same limit $V'$ on $\Delta$. Obviously $\mu_\Delta(U \wedge V_k') \in HD$ and $\lim_{k\to\infty} \mu_\Delta(U \wedge V_k') = \mu_\Delta(V') \leqq \mu_\Delta(V) = u$ on $F$. Since $u$ is $\underline{HD}$-minimal on $F$, we have $\mu_\Delta(V') = cu$ where $c$ is a constant. Consequently $V' = c\lambda_\Delta(u) = cV$. Thus, it is proved that $\lambda_\Delta(u)$ is $\underline{HD}$-minimal on $\Delta$.

One of the most important results of CONSTANTINESCU-CORNEA ([1], Satz 10, p. 211) is stated as follows: If $\gamma$ is a maximal $HD$-indivisible set, then $\omega(p, \gamma, F)$ is an $\underline{HD}$-minimal function on $F$. Conversely, if $u$ is an $\underline{HD}$-minimal function on $F$, then $u = c\omega(p, \gamma, F)$ and $\gamma$ is a maximal $HD$-indivisible set. Let us denote by $O_{HD_n}$ $(1 \leqq n \leqq \infty)$ the class of Riemann surfaces $F$ such that the set $\Gamma(\mathcal{M})$ of all the points of the ideal boundary of $F$ consists of at most $n$ maximal $HD$-indivisible sets.

Using Lemma 3, we can prove an extension of Theorem 1.

**Theorem 10.** *Let $F$ be a Riemann surface belonging to $O_{HD_n}$ $(1 \leqq n \leqq \infty)$ and $\Delta$ be a subregion on $F$ which does not belong to $SO_{HD}$. Then $\Delta$ belongs to $O_{AD}$*[1]. (CONSTANTINESCU-CORNEA [1], KURAMOCHI [4]).

*Proof.* Suppose that the set $\Gamma(\mathcal{M})$ on the ideal boundary of $F$ consists of just $m$ $(\leqq n)$ maximal $HD$-indivisible sets $\gamma_k$ $(k = 1, 2, \ldots, m)$. Let $\omega_k$ $(k = 1, 2, \ldots, m)$ be the harmonic measure of $\gamma_k$ with respect to $F$. Then, by the fact just stated above, $\omega_k$ is $\underline{HD}$-minimal on $F$. Since $\Delta \notin SO_{HD} = SO_{HBD}$, there exists a positive bounded harmonic function $U$ having a finite Dirichlet integral on $\Delta$ and vanishing continuously on the relative boundary $\beta$ of $\Delta$. Since $\mu_\Delta(U)$ is Dirichlet-bounded on $F$, we can write

$$\mu_\Delta(U) = \sum_{k=1}^{m} c_k \omega_k .$$

Since $\mu_\Delta(U) > 0$, there exists at least one $k$ such that $c_k \omega_k > 0$. For such a $k$, $\mu_\Delta\left(\frac{1}{c_k} U\right) \geqq \omega_k$ on $F$. By Lemma 3, we see that $\lambda_\Delta(\omega_k)$ is $\underline{HD}$-minimal on $\Delta$. Let $p = \varphi(\zeta)$ be the univalent conformal mapping of the disc $|\zeta| < 1$

[1] The following proof is due to MATSUMOTO [2].

onto the universal covering surface $\tilde{\Delta}$ of $\Delta$. Denote by $E_1$ the set of all the points $\zeta = e^{i\vartheta}$ for which $\lambda_\Delta(\omega_k(\varphi(\zeta)))$ has angular limit 1. Then, $E_1$ is of positive measure and the image of a certain $HD$-indivisible set $\gamma_k$ on the ideal boundary of $\Delta$. Consider now any single-valued Dirichlet-bounded analytic function $w = f(p)$ on $\Delta$. Then the function $f(\varphi(\zeta))$ has the same angular limit almost everywhere on $E_1$, and therefore $f(\varphi(\zeta))$ must be identically constant.

Applying this theorem, we obtain an extension of a result of KURAMOCHI [4].

**Theorem 11.** *Let $F$ be a Riemann surface belonging to $O_{HD_n}$ $(1 \leqq n \leqq \infty)$ and $w = f(p)$ be a non-constant single-valued meromorphic function on $F$. Let $G$ be a disc $|w - w_0| < \varrho$ on the $w$-plane and $\Delta$ be a connected component of $f^{-1}(G)$ and $\Phi_\Delta$ be the corresponding covering surface of $G$. Then, the mapping $f_\Delta$ (the restriction of $f$ to $\Delta$) of $\Delta$ into $G$ is of type-Bl, provided that the area of $\Phi_\Delta$ is finite.* (MATSUMOTO [2][1]).

*Proof.* Contrary to the assertion, suppose that $f_\Delta$ is not of type-Bl. Then, by Theorem 5, there exists a disc $G_1$: $|w - w_0| < \varrho_1$ $(\varrho_1 < \varrho)$ such that at least one component $\Delta_1$ of $f_\Delta^{-1}(G_1)$ does not belong to $SO_{HB}$. Let $\mathscr{G}_G(w; w_0) = \log \left|\frac{\varrho}{w - w_0}\right|$. Then $\mathscr{G}_G(w; w_0) = \log \frac{\varrho}{\varrho_1} = A > 0$ on $|w - w_0| = \varrho_1$. Consider the superharmonic function

$$\Omega(w) = \min(\mathscr{G}_G(w; w_0), A) \quad \text{on } G. \tag{8}$$

Then, by an elementary calculation, we see that the superharmonic function $\Omega(f_\Delta(p))$ has a finite Dirichlet integral on $\Delta$ [2]. By the Dirichlet principle, the greatest harmonic minorant $u$ of $\Omega(f_\Delta)$ is Dirichlet-bounded on $\Delta$. Now, let $\omega = \omega(p, \Delta_1)$ be the harmonic measure of the ideal boundary of $\Delta_1$; then $\omega > 0$ in $\Delta_1$. We denote by $\omega^*$ the standard subharmonic extension of $\omega$ to $\Delta$ such that $\omega^* = \omega$ in $\Delta_1$ and $\omega^* = 0$ in $\Delta - \Delta_1$. Obviously

$$A\omega^* \leqq \Omega(f_\Delta) \quad \text{in } \Delta. \tag{9}$$

Let $v$ be the least harmonic majorant of $A\omega^*$ in $\Delta$. Then, by (9), we have $0 < v \leqq u$ in $\Delta$. Accordingly, $\Delta$ cannot belong to $SO_{HD}$. This contradicts Theorem 10.

5. We close this chapter with some remarks. OHTSUKA ([10], [14]) and KURAMOCHI [5] have tried to extend the theory of cluster sets stated in II to the case of open Riemann surfaces. In the study of cluster sets and, more generally, boundary behaviours of functions analytic or harmonic on open Riemann surfaces, it seems very important to introduce appropriate compactifications of open Riemann surfaces (as to the compactification,

[1] For an earlier related theorem, see MORI [3].

[2] Practically we have $D(\Omega(f_\Delta)) \leqq \frac{1}{\varrho_1} D(f)$.

see ROYDEN [4])[1]. We note that, to this direction, recently many important contributions have been made by BRELOT [3—7], NAÏM [1], PARREAU [3], KAKUTANI [2], DOOB [5, 6], SARIO [12] and SAVAGE [1] from various points of view.

# Appendix: Cluster sets of pseudo-analytic functions[2]

**1. Differentiable quasiconformal mappings**[3]. Let

$$w = T(z) = u(x, y) + iv(x, y)$$

be a topological mapping of a domain $D$ in the $z$-plane onto a domain $\Delta$ in the $w$-plane. Suppose that $w = T(z)$ belongs to class $C^1$ and its Jacobian $J(z) = u_x v_y - u_y v_x$ is positive at every point $z$ in $D$. Then, any infinitesimal circle with center $z$ and of radius $\varepsilon$ is transformed into an infinitesimal ellipse with center $w = T(z)$ and of the major axis $a\varepsilon$ and the minor $b\varepsilon$. If the dilatation $d(z) = a/b$ is bounded in $D$:

$$d(z) = a/b \leqq K, \tag{1}$$

where $K$ is a constant $\geqq 1$, then $w = T(z)$ is called a *quasiconformal mapping of parameter* $K$.

The condition (1) is equivalent to each of the following conditions[4]:

$(\alpha)$ $\max\limits_{0 \leqq \theta \leqq 2\pi} |w_x \cos\theta + w_y \sin\theta|^2 = \max\limits_{\theta} \frac{|dw|^2}{|dz|^2} \leqq K J(z)$ ;

$(\beta)$ $u_x^2 + u_y^2 + v_x^2 + v_y^2 \leqq \left(K + \frac{1}{K}\right) J(z)$ ;

$(\gamma)$ $\left|\frac{w_x + i w_y}{w_x - i w_y}\right| \leqq \frac{K-1}{K+1}$

(BERS [1]).

[1] As to the topology introduced by MARTIN [1], see a nice description of PARREAU [1].

[2] As to the recent development of the theory of quasiconformal mappings, see KÜNZI [1].

[3] The concept of quasiconformality is due to GRÖTZSCH [1, 2, 3].

[4] Introducing complex derivatives

$$p = \frac{\partial w}{\partial z} = \frac{1}{2}(w_x - i w_y), \quad q = \frac{\partial w}{\partial \bar z} = \frac{1}{2}(w_x + i w_y),$$

we have

$$(|p| - |q|)\,|dz| \leqq |dw| \leqq (|p| + |q|)\,|dz|$$

and

$$|p|^2 - |q|^2 = J(z).$$

Hence

$$d(z) = \frac{|p| + |q|}{|p| - |q|},$$

$$2(|p|^2 + |q|^2) = d(z) + 1/d(z) \quad \text{and} \quad \frac{|q|}{|p|} = \frac{d(z) - 1}{d(z) + 1}.$$

Using these relations, we can easily prove the required equivalencies. The definition of quasiconformality stated in terms of complex derivatives is due to AHLFORS [8].

**2. Geometric definition of quasiconformality.** Let $Q$ be a simply connected domain in the $z$-plane bounded by a Jordan curve, and let $z_1, z_2, z_3, z_4$ be four distinct boundary points of $Q$, which lie in this order on the positively oriented boundary curve. We call such a configuration a *quadrilateral* and denote it by $Q(z_1, z_2, z_3, z_4)$. Map the domain $Q$ conformally onto a rectangle: $0 < \xi < M$, $0 < \eta < 1$, in the $\zeta$-plane ($\zeta = \xi + i\eta$), in such a manner that $z_1, z_2, z_3, z_4$ correspond to the vertices $\zeta = 0, M, M + i, i$ respectively. We call the positive number $M$ the *modulus* (or *module*) of the quadrilateral $Q(z_1, z_2, z_3, z_4)$ or simply $\operatorname{mod} Q$. The following geometric definition is due to PFLUGER [3], AHLFORS [8] and MORI [9].

*Definition.* A sense-preserving topological mapping $w = T(z)$ of a plane domain $D$ onto another such domain $\Delta$ is called *K-quasiconformal* or, simply, a *KQC mapping*, provided that for any quadrilateral $Q$ contained in $D$ with its boundary,

$$\operatorname{mod} T(Q) \leqq K \operatorname{mod} Q\,, \tag{2}$$

where $T(Q)$ is the image of $Q$ and $K$ is a constant $\geqq 1$.

Since $\operatorname{mod} Q(z_2, z_3, z_4, z_1) = 1/\operatorname{mod} Q(z_1, z_2, z_3, z_4)$, the condition (2) can be written in the form

$$K^{-1} \operatorname{mod} Q \leqq \operatorname{mod} T(Q) \leqq K \operatorname{mod} Q\,; \tag{2'}$$

hence, *if $w = T(z)$ is K-quasiconformal, then its inverse is also K-quasiconformal.* It is obvious that *the composition of a KQC mapping and a conformal mapping is a KQC mapping.*

**3.** Starting from the geometric definition of quasiconformality in the sense of PFLUGER-AHLFORS-MORI, we can deduce some important analytic properties.

**Theorem 1.** *Let $w = T(z) = u(x, y) + iv(x, y)$ be a K-quasiconformal mapping* (in the sense of PFLUGER-AHLFORS-MORI) *of a plane domain $D$ in the z-plane ($z = x + iy$) onto a domain $\Delta$ in the w-plane ($w = u + iv$). Then*

(i) *$w = T(z)$ is totally differentiable almost everywhere in $D$* (MORI [9])[1];

(ii) *for each point $z$, at which $w = T(z)$ is totally differentiable,*

$$\max_{\theta} \frac{|dw|^2}{|dz|^2} \leqq K\,J(z)$$

(MORI [9]);

(iii) *for almost every $y = y_0$, $w = T(x, y_0)$ is absolutely continuous in $x$ on any closed interval in the intersection of $y = y_0$ and $D$* (MORI [9], STREBEL [1], PFLUGER [5])[2];

[1] To prove (i), MORI uses Rademacher-Stepanoff's theorem (SAKS [1], p. 310).

[2] The proofs of STREBEL [1] and PFLUGER [5] for (iii) are simpler than that of MORI [9]. From (iii) and (iv), it follows that $w = T(z)$ is absolutely continuous in the sense of TONELLI in $D$ (SAKS [1], p. 169).

(iv) *the partial derivatives of* $w = T(z)$ *are locally square integrable in* $D$ (BERS [1]);

(v) $w = T(z)$ *is a measurable mapping; more precisely, for every measurable set* $e \subset D$, *the image* $T(e)$ *is measurable and has measure*

$$\iint_e J(z)\,dx\,dy$$

(MORREY [1], JENKINS [2])[1].

In virtue of these properties (i)—(v), we can extend almost all known results on differentiable quasiconformal mappings to the class of $K$-quasiconformal mappings (in the sense of PFLUGER-AHLFORS-MORI), by simple recapitulation of the original proofs.

As an example, we prove

**Theorem 2.** *If* $w - T(z)$ *be a topological mapping of a plane domain* $D$ *onto another such domain. If* $w = T(z)$ *is* $KQC$ *in a neighborhood of each point of* $D$, *then it is* $KQC$ *in* $D$ (AHLFORS [8], MORI [9])[2].

*Proof*[3]. Let $Q$ be a quadrilateral whose closure is contained in $D$. Making auxiliary conformal mappings, we may assume that $Q$ is a rectangle $0 < x < M$, $0 < y < 1$ and the image $T(Q)$ is $0 < u < M'$, $0 < v < 1$, and $z = 0, M, M + i, i$ correspond to $w = 0, M', M' + i, i$. Since the rectangle $Q$ is covered by countably many neighborhoods, in each of which $w = T(z)$ is $KQC$, we see, by (iii), that for almost all $0 < y_0 < 1$, the length of the $T$-image of the segment $0 < x < M$, $y = y_0$ is represented by the integral $\int_0^M |dT(x, y_0)/dx|\,dx \leqq +\infty$. Since it connects a point on $u = 0$ to a point on $u = M'$, we have $M' \leqq \int_0^M |dT(x, y_0)/dx|\,dx$ for almost all $0 < y_0 < 1$. By Schwarz's inequality, $M^{-1}M'^2 \leqq \int_0^M |dT(x, y_0)/dx|^2 dx$ and by Fubini's theorem, $M^{-1}M'^2 \leqq \int_0^1 \left[\int_0^M |dT(x, y)/dx|^2 dx\right] dy = \iint_Q |dT(x, y)/dx|^2 dx\,dy$. By (ii) and (v), we have

$$M^{-1}M'^2 \leqq K \iint_Q J(z)\,dx\,dy = KM' \text{ ; i. e.,}$$

$\operatorname{mod} T(Q) \leqq K \operatorname{mod} Q$.

Similarly, we can prove

**Theorem 3.** *Let* $w = T(z)$ *be a* $KQC$ *mapping of* $|z| < 1$ *onto* $|w| < 1$. *Then,* $w = T(z)$ *can be extended to a topological mapping of the closed disc* $|z| \leqq 1$ *onto* $|w| \leqq 1$ (AHLFORS [8], MORI [9])[4].

[1] AHLFORS gave a simple proof for (v) in his lecture at Osaka University in 1955.

[2] This theorem means that quasiconformality is a local property.

[3] The proof is due to MORI [9].

[4] Cf. SAKAI [1], YÛJÔBÔ [4, 5].

Hence, by reflections with respect to $|z| = 1$ and $|w| = 1$, we can extend $w = T(z)$ to a $KQC$ mapping of the whole $z$-sphere onto the whole $w$-sphere, since the unit circumference $|z| = 1$ is *deletable* for the extended topological mapping $w = T(z)$ by Ahlfors' theorem[1].

**Theorem 4.** *Let $w = T(z)$ be a $KQC$ mapping of $|z| < 1$ onto $|w| < 1$ such that $T(1) = 1$. If the point $z$ tends to 1 inside a Stolz angle $|\arg(1 - z)| < \varphi < \pi/2$, its image $w = T(z)$ tends to 1 also inside a Stolz angle $|\arg(1 - w)| < \psi < \pi/2$, where $\psi$ depends only on $\varphi$ and $K$* (Mori [9], Jenkins [2]).

*Proof.* It is sufficient to prove the following fact: Let $w = T(z)$ be a $KQC$ mapping of the half disc $D$: $|z| < 1$, $\mathfrak{I}(z) > 0$ onto another half disc $\Delta$: $|w| < 1$, $\mathfrak{I}(w) > 0$ such that $z = -1, 0, 1$ correspond to $w = -1, 0, 1$ respectively. Then, the $T$-image of any circular arc $\gamma$ in $D$, passing through $z = -1$ and $z = 1$, lies above a certain circular arc $\Gamma$ in $\Delta$ passing through $w = -1$ and $w = 1$. By reflections with respect to the real axes, we can extend $w = T(z)$ to be a $KQC$ mapping of $|z| < 1$ onto $|w| < 1$. Let $z_0$ be a point in the interval $(-1, 1)$ of the real axis and $w_0 = T(z_0)$ be its image. Making auxiliary conformal mappings

$$Z = \frac{z - z_0}{1 - z_0 z}, \qquad W = \frac{w - w_0}{1 - w_0 w},$$

we obtain a $KQC$ mapping $W = W(Z)$ of $|Z| < 1$ onto $|W| < 1$ with $W(0) = 0$. Hence, by an extension of Schwarz's lemma[2], we have

$$4^{-K}|Z|^K \leqq |W(Z)| \; ; \quad \text{i. e.}$$

$$\left|\frac{w - w_0}{1 - w_0 w}\right| \geqq 4^{-K}\left|\frac{z - z_0}{1 - z_0 z}\right|^K.$$

Consequently, if $\varrho$ is the distance of $\gamma$ from $z = 0$, then the $T$-image of $\gamma$ lies above the circular arc $\Gamma$ in $\Delta$, passing through $w = -1$ and $w = 1$, with the distance $4^{-K}\varrho^K$ from $w = 0$.[3]

**4.** Let $F$ and $\Phi$ be two open abstract Riemann surfaces. Let $q = T(p)$ be a sense-preserving topological mapping of $F$ onto $\Phi$ which is $K$-quasiconformal in some neighborhood of each point $p$ of $F$. Then, $T$ is called a

---

[1] Cf. Ahlfors [8]; Theorem 4, p. 9. This result of Ahlfors has been extended by Mori [9] and Strebel [1] independently in the following way: Let $D$ be a plane domain, and $E$ be a relatively closed set in $D$ which consists of at most countably many sets of finite (outer) linear measure. Then, a topological mapping $w = T(z)$ of $D$, which is $KQC$ in some neighborhood of each point of $D - E$, is $KQC$ in $D$.

[2] Let $w = T(z)$ be a $KQC$ mapping of $|z| < 1$ onto $|w| < 1$ such that $T(0) = 0$. Then, for each $0 < |z| < 1$,

$$4^{-K}|z|^K \leqq |T(z)| \leqq 4\,|z|^{K^{-1}}.$$

Cf. Mori [9], Hersch-Pfluger [1].

[3] Cf. Carathéodory [1], p. 53.

$KQC$ mapping of $F$ onto $\Phi$. Let $G$ be an open set (not necessarily connected) whose closure $\overline{G}$ is compact on $F$. We assume that the boundary $\gamma$ of $G$ consists of a finite number of simple closed curves. We divide the boundary $\gamma$ into two disjoint sets $\gamma_0$ and $\gamma_1$. Let $\omega$ be the harmonic measure in $G$ with boundary values 0 on $\gamma_0$ and 1 on $\gamma_1$. The (harmonic) modulus (in the sense of SARIO-PFLUGER) of the generalized annulus $G = \{\gamma_0, \gamma_1\}$ is defined as $2\pi/D(\omega)$, where $D(\omega)$ denotes the Dirichlet-integral of $\omega$ on $G$ (cf. Paragraph 2, § 1, IV). In virtue of Theorem 1, we can prove that if $q = T(p)$ is a $KQC$ mapping of $F$ onto $\Phi$, then

$$K^{-1} \operatorname{mod} G \leqq \operatorname{mod} T(G) \leqq K \operatorname{mod} G$$

for any (relatively compact) annulus $G$ on $F$.

Hence, applying the modular criterion (Theorem 2, § 1, IV), we can prove

**Theorem 5.** *Let $q = T(p)$ be a $KQC$ mapping of an open Riemann surface $F$ onto another Riemann surface $\Phi$. Then, if $F$ belongs to $O_G$, $\Phi$ also belongs to $O_G$* (PFLUGER [1]).

**Corollary.** *Let $w = T(z)$ be a $KQC$ mapping of $|z| < 1$ onto $|w| < 1$. Let $E_z$ be a point set on the closed disc $|z| \leqq 1$, and $E_w$ its image on $|w| \leqq 1$. Then, if $E_z$ is of inner or outer (logarithmic) capacity zero, $E_w$ has the same property* (cf. MORI [9])[1].

For the later purpose, we shall give an alternative proof for Theorem 5 based on the Gross property (cf. Theorem 3, § 1, IV).

*Proof of Theorem* 5. Assume, contrary to the assertion, that $\Phi$ is of hyperbolic type. Without loss of generality, we may assume that $\Phi$ is a covering Riemann surface of the $w$-sphere. Let $g(w, w_0)$ be the Green function on $\Phi$ with pole at $w_0$ and $h(w, w_0)$ be its conjugate. Then the composed function

$$\zeta(w) = e^{-g(w, w_0) - ih(w, w_0)}$$

is a many-valued analytic function on $\Phi$ and any branch of $\zeta(w)$ has a simple zero at $w_0$. Let $w(\zeta) = e(\zeta, 0)$ be any inverse functional element, at $\zeta = 0$, of $\zeta(w)$. We continue $e(\zeta, 0)$ analytically along every radius of the unit disc $|\zeta| < 1$, with rational character, and define a starshaped domain $\Delta_\zeta$ in the sense of GROSS lying in $|\zeta| < 1$. We denote again by $w = w(\zeta)$ the single-valued meromorphic function which is determined by the element $e(\zeta, 0)$ in the simply connected domain $\Delta_\zeta$. In virtue of Brelot-Choquet's theorem [1], for almost every $e^{i\vartheta}$ the radius $\varrho_\vartheta$: $\zeta = re^{i\vartheta}$, $0 \leqq r < 1$ is

[1] By definition, the inner capacity of a set is the supremum of capacities of closed sets contained in it and the outer capacity of a set is the infimum of the inner capacities of open sets containing it. Hence, to prove the corollary, we may assume that $E_z$ and $E_w$ are closed. We can extend $w = T(z)$ to be a $KQC$ mapping of the complement $\mathscr{C}E_z$ with respect to the $z$-plane onto $\mathscr{C}E_w$. Then the assertion follows immediately from Theorem 5.

contained in $\Delta_\zeta$. Consider the image $\Delta_w$ of $\Delta_\zeta$ by $w = w(\zeta)$ and then the counter-image $\Delta = T^{-1}(\Delta_w)$ on $F$. Obviously, the composed mapping $p = p(\zeta)$ is a $KQC$ mapping of $\Delta_\zeta$ onto $\Delta$. We note that for almost every $e^{i\vartheta}$, the image of the radius $\varrho_\vartheta$ by the mapping $p = p(\zeta)$ is a path in $\Delta$ starting from the point $p_0 = T^{-1}(w_0)$ and terminating at the ideal boundary of $\Delta$. By using the property (ii) of $KQC$ mapping in Theorem 1 and the recapitulation of the proof of Theorem 3, § 1, IV, we arrive at a contradiction.

Let $F$ be an open Riemann surface and $w = f(p)$ be a single-valued complex-valued function defined on $F$. If $w = f(p)$ is an interior transformation on $F$ in the sense of STOÏLOW [9], then all the elements $q = [p, f(p)]$ form a covering Riemann surface $\Phi$ of the $w$-plane. We call $w = f(p)$ *a K-pseudo-analytic function* or simply *a KPA function* on $F$, provided that either $w = f(p)$ is identically equal to a constant or the mapping $\varphi\colon p \to q = [p, f(p)]$ is a $KQC$ mapping of $F$ onto $\Phi$. Let $\Psi$ be another Riemann surface conformally equivalent to $\Phi$ and let $q = \psi(r)$ be a conformal mapping of $\Psi$ onto $\Phi$. Then, the composed mapping $T$: $p \to r$ is a $KQC$ mapping of $F$ onto $\Psi$. Thus, we obtain the following representation:

$$w = f(p) \equiv \phi[T(p)], \tag{3}$$

where $w = \phi(r) \equiv f(\varphi^{-1}(\psi(r)))$ is a single-valued analytic function on $\Psi$.

We prove an extension of Beurling's theorem [2].

**Theorem 6.** *Let $w = f(z)$ be a KPA function in $|z| < 1$, and suppose that the Riemann covering surface $\Phi$ generated by $w = f(z)$ of the $w$-plane has a finite spherical area. Then, except for a set of outer capacity zero on $|z| = 1$, $w = f(z)$ possesses angular limits. Further, if $w = \alpha$ is an ordinary value for $f(z)$ in the sense of* BEURLING[1], *the set of points on $|z| = 1$, where $f(z)$ has the angular limit $\alpha$, is of outer capacity zero* (MORI [9], LOHWATER [8], JENKINS [2]).

*Proof*[2]. By (3), $w = f(z)$ is written in the form:

$$w = f(z) = \phi[T(z)],$$

where $\zeta = T(z)$ is a $KQC$ mapping of $|z| < 1$ onto $|\zeta| < 1$ and $w = \phi(\zeta)$ is analytic in $|\zeta| < 1$. For the analytic function $w = \phi(\zeta)$, Beurling's theorem holds. Let $E_\zeta$ be the exceptional set on $|\zeta| = 1$, at each point of which $\phi(\zeta)$ has no angular limit, and let $E_z$ be the image of $E_\zeta$ by $z = T^{-1}(\zeta)$ (cf. Theorem 3). Since $E_\zeta$ is of outer capacity zero, so is $E_z$ by Corollary of Theorem 5. By applying Theorem 4, we see that $w = f(z)$ possesses an angular limit at every point $e^{i\theta} \notin E_z$. The remaining part is proved in the same way.

[1] Let $s(\varrho)$ denote the spherical area of the part of $\Phi$ above the disc $|w - \alpha| < \varrho$. If $\varliminf_{\varrho \to 0} \frac{s(\varrho)}{\pi \varrho^2} < \infty$, then $\alpha$ is called an ordinary value in the sense of BEURLING [2].

[2] The proof is due to MORI [9].

**5. Cluster sets of pseudo-analytic functions.** At the beginning, we prove

**Theorem 7.** *Let $E$ be a compact set of capacity zero, $D_1$ a domain containing $E$ completely in its interior and $D = D_1 - E$. Suppose that $w = f(z)$ is a $KPA$ function in $D$. Then, the cluster set $C_D(f, z_0)$ at each point $z_0$ of $E$ is either a single point or the whole $w$-plane* (OHTSUKA [14])[1].

*Proof.* Without loss of generality, we may assume that $D_1$ is the unit disc: $|z| < 1$. Let $\Phi$ be the Riemann covering surface, generated by $w = f(z)$ on $D$, of the $w$-plane. Then, by the uniformization theory, we can map $\Phi$ conformally onto a plane domain $\Delta$ in the $\zeta$-plane. We may assume that the boundary component of $\Delta$, which corresponds to $|z| = 1$, is $|\zeta| = 1$ and $\Delta$ is contained in the unit disc $|\zeta| < 1$. Denote by $w = \phi(\zeta)$ the function which maps $\Delta$ conformally onto $\Phi$. Then, by (3), we have $w = f(z) = \phi[T(z)]$ where $\zeta = T(z)$ is a $KQC$ mapping of $D$ onto $\Delta$ such that the boundary component $|z| = 1$ of $D$ corresponds to $|\zeta| = 1$. By reflections with respect to $|z| = 1$ and $|\zeta| = 1$, we can extend $T$ to be a $KQC$ mapping of the double $\hat{D}$ onto the double $\hat{\Delta}$. Since $\hat{D}$ is of parabolic type, $\hat{\Delta}$ is also of parabolic type. Hence the complement $E_\zeta$ of $\Delta$ with respect to $|\zeta| < 1$ is a compact set of capacity zero. Obviously, we can continue $w = T(z)$ to be a topological mapping of $|z| < 1$ onto $|\zeta| < 1$[2]. By Theorem 1, § 2, III, the cluster set $C_\Delta(\phi, \zeta_0)$ at each point $\zeta_0$ of $E_\zeta$ is either a single point or the whole $w$-plane. From this fact, our assertion follows immediately.

Theorem 7 shows the possibility of extending the results on cluster sets stated in II to the case of pseudo-analytic functions.

Let $G$ be a circular disc: $|w - w_0| < \varrho$ and $e_w$ be a totally disconnected compact set of positive capacity in the interior of $G$. Let $\omega$ be the harmonic measure in $G - e_w$ with boundary values 0 on $|w - w_0| = \varrho$ and 1 on $e_w$, $\tilde{\omega}$ its conjugate and $\int_{|w - w_0| = \varrho} d\tilde{\omega} = d$. We consider the function[3]

$$\zeta(w) = \exp\left\{-\frac{2\pi}{d}(\omega + i\tilde{\omega})\right\} \tag{4}$$

[1] An analogue of Theorem 1, § 2, II.

[2] More precisely, since a compact set of capacity zero is of linear measure zero, we can extend $T$ to be a $KQC$ mapping of $|z| < 1$ onto $|\zeta| < 1$ (see foot-note 1 p. 112).

[3] Any branch $w = w(\zeta)$ of the inverse of $\zeta = \zeta(w)$ is single-valued and regular on $|\zeta| = 1$. Continue $w = w(\zeta)$ analytically, with rational characters, along every radius starting from the point on $|\zeta| = 1$ toward $\zeta = 0$. Then, for almost every $\vartheta$, the continuation along the radial segment $S_\vartheta$: $\zeta = |\zeta| e^{i\vartheta}$, $e^{-\frac{2\pi}{d}} < |\zeta| < 1$ is possible. Thus, $w = w(\zeta)$ defines a single-valued univalent analytic function $w = \psi(\zeta)$ in the annulus $e^{-\frac{2\pi}{d}} < |\zeta| < 1$ with radial slits whose arguments $\vartheta$ in the interval $(0, 2\pi)$ form a set of linear measure zero. This analogue of Brelot-Choquet's theorem can be proved directly, since the irregular points for the Dirichlet problem form a set of type $F_\sigma$ of capacity zero.

in the domain $G_1 = G - e_w$ and introduce the conformal metric

$$d s_w = |\zeta'(w)|\,|d w| \tag{5}$$

in $G_1$.

**Lemma.** *Let $F$ be a finite covering surface of the basic surface $G_1 = G - e_w$. Then, using the conformal metric* (5), *we have*

$$A \leqq hL, \tag{6}$$

*where $A$ is the area of $F$, $L$ the total length of the relative boundary $\beta$ of $F$ and $h$ is a positive constant independent of $F$* (T. Yosida [2]).

*Proof.* By reflection with respect to the circle: $|w - w_0| = \varrho$, we construct the double $\hat{F}$ of $F$ above the double $\hat{G}_1$ of $G_1 = G - e_w$. Let $\partial\hat{F}$ denote the boundary of $\hat{F}$. We denote by $A_s$ and $L_s$ the spherical area of $F$ and the spherical length of $\beta$ with respect to the metric (5). Then, by a well-known formula,

$$\begin{aligned} 2A_s &= \iint\limits_{\hat{F}} \frac{|\zeta'(w)|^2}{(1+|\zeta(w)|^2)^2}\,du\,dv\,, \quad (w = u + iv)\,, \\ &\leqq \frac{1}{2}\int\limits_{\partial\hat{F}} \frac{|\zeta(w)|}{1+|\zeta(w)|^2}\,|\zeta'(w)|\,|dw| \\ &\leqq \frac{h_1}{2}\int\limits_{\partial\hat{F}} \frac{|\zeta'(w)|}{1+|\zeta(w)|^2}\,|dw| = 2\cdot\frac{h_1}{2}\int\limits_{\beta} \frac{|\zeta'(w)|\,|dw|}{1+|\zeta(w)|^2} = h_1 L_s\,, \end{aligned}$$

since $|\zeta(w)|$ is bounded: $|\zeta(w)| \leqq h_1$ in $\hat{G}_1$ for a certain positive number $h_1$; whence follows that $A_s \leqq \frac{1}{2} h_1 L_s$ and so $A \leqq hL$ where $h = 2h_1$.

**Theorem 8.** *Let $D$ be an arbitrary domain, $\Gamma$ its boundary, $E$ a compact set of capacity zero on $\Gamma$ and $z_0$ a point of $E$. Suppose that $w = f(z)$ is a KPA function in $D$. Then*

(i) *if $\alpha \in C_D(f, z_0) - C_{\Gamma - E}(f, z_0)$ is an exceptional value of $w = f(z)$ in a neighborhood of $z_0$, then either $\alpha$ is an asymptotic value of $f(z)$ at $z_0$ or there is a sequence of accessible boundary points $z_n \in E$ $(n = 1, 2, \ldots)$ converging to $z_0$ such that $\alpha$ is an asymptotic value of $f(z)$ at each point $z_n$* (Noshiro [8]);

(ii) *if $z_0$ is an accumulation point of $\Gamma - E$, i. e. $z_0 \in \overline{(\Gamma - E)}$, then $\Omega = C_D(f, z_0) - C_{\Gamma - E}(f, z_0)$ is an open set* (Yosida [2], Ohtsuka [10]);

(iii) *if $z_0 \in \overline{(\Gamma - E)}$ and if $\Omega$ is not empty, then $\Omega - R_D(f, z_0)$ is at most of capacity zero* (Yosida [2], Ohtsuka [10]);

(iv) *if $D$ is a simply connected domain of hyperbolic type, and if $\Omega$ is not empty, and further if $f(z)$ is bounded in the intersection of $D$ and some neighborhood of $z_0$, then $w = f(z)$ takes every value, with one possible exception, belonging to each component $\Omega_n$ of $\Omega$, infinitely often in any neighborhood of $z_0$;*

(v) *if each point of $E$ belongs to a non-degenerate continuum disjoint from $D$ and if $\Omega$ is not empty, then $w = f(z)$ assumes every value, with two possible exceptions, belonging to each component $\Omega_n$ of $\Omega$, infinitely often in any neighborhood of $z_0$* [1].

*Proof.* First we note analytic properties of $KQC$ mappings in Theorem 1 and the fact that Ahlfors' theory of covering surfaces is also useful in the case of $KPA$ functions. We can prove (i), (iv), (v) by simple recapitulation of the proofs of corresponding theorems in the case of analytic functions. To prove (ii), we follow the entirely similar argument used in the proof of Theorem 4, § 4, II, and arrive at the stage to consider the case in which $w_0 \notin C_{\Gamma-E}(f, z_0)$ is an asymptotic value of $w = f(z)$ along a path $\Lambda$ in $D$ terminating at $z_0$. For a sufficiently small positive number $\varrho$, we denote by $\Delta$ the component of the counter-image $f^{-1}(G)$ of $G$: $|w - w_0| < \varrho$ which contains the last part of the path $\Lambda$. We may assume that the boundary $\gamma$ of $\Delta$ consists of a compact subset $e$ of $E$ and at most countably many simple curves (boundary relative to $D$). The Riemann covering surface $\Phi_\Delta$, generated by $w = f(z)$ on $\Delta$, of the basic surface $G$ has Gross' property and so Iversen's property on $G$. Hence, by Stoïlow's principle (cf. Paragraph 1, § 2, IV), either $\Phi_\Delta$ covers every point $w$ the same finite number of times except for a totally disconnected (relatively) closed set which consists of the asymptotic values at some points of $e$; or the cluster set $C_\Delta(f)$ (in the large) is $\overline{G}$. However, since $z_0 \in \overline{(\Gamma - E)}$, $C_\Delta(f, z_0)$ contains a continuum which connects $w_0$ to a point of $|w - w_0| = \varrho$. Consequently, $C_\Delta(f) = \overline{G}$ and $\Phi_\Delta$ has an infinite area, by Gross' property. We shall show that $C_\Delta(f, z_0)$ is also $\overline{G}$. For this purpose, it suffices to prove that $G - R_\Delta(f, z_0)$ is of capacity zero. Otherwise, there would exist a positive number $r_0$, such that there is no point of $e$ on $|z - z_0| = r_0$, and a totally disconnected compact set $e_w$ of positive capacity lying completely in $G$ such that $w_0 \notin e_w$ and the set of values of $w = f(z)$ in the intersection $\Delta_{r_0}$ of $\Delta$ with the disc $|z - z_0| < r_0$ contains no points of $e_w$. The open set $\Delta_{r_0}$ consists of a finite number of connected domains. Hence, the Riemannian image $\Phi_{r_0}$ of $\Delta_{r_0}$ by $w = f(z)$ consists of the same number of open covering surfaces of the basic surface $G_1 = G - e_w$. Introducing the conformal metric (5) in $G_1$, we see that the area of $\Phi_{r_0}$ is infinite, since the part of $\Phi_{r_0}$ above a small disc $|w - w_0| < \varrho_1$ has an infinite area, by the fact just stated above. Now, with the aid of the level curve $\Gamma_\lambda$: $\varrho(z) = \text{const.} = \lambda$ $(0 < \lambda < \infty)$ of an Evans-Selberg's function $\zeta = \chi(z) = \varrho(z)\, e^{i\,v(z)}$ (cf. § 1, II) associated with the set $e$, we denote by $\Delta(\lambda)$ the common part of $\Delta_{r_0}$ and the domain exterior to $\Gamma_\lambda$, by $A(\lambda)$ the area of the Riemannian image, by $w = f(z)$, of $\Delta(\lambda)$ and by $L(\lambda)$ the total length

[1] (i), (ii), (iii), (iv), (v) are extensions of Theorem 1, § 3, II; Theorem 4, § 4, II; Theorem 5, § 4, II; Theorem 10, § 4, II; Theorem 2, § 5, II respectively. As to further extensions of (ii) and (iii), cf. OHTSUKA [14].

of the image of the intersection of $\Gamma_\lambda$ and $\Delta_{r_0}$ with respect to the metric (5). Then, $\Phi_{r_0}$ is regularly exhaustible in the sense of AHLFORS; i. e.

$$\varliminf_{\lambda\to\infty} L(\lambda)/A(\lambda) = 0$$

(cf. the proof of Theorem 6, § 4, II). On the other hand, by the preceding lemma, $A(\lambda) \leqq h(L(\lambda) + L_0)$, where $L_0$ denotes the total length of the image of the intersection of $|z - z_0| = r_0$ and $\Delta$; hence $\varliminf_{\lambda\to\infty} L(\lambda)/A(\lambda) \geqq \geqq 1/h > 0$. Contradiction.

To prove (iii), we follow the similar argument used in the proof of Theorem 5, § 4, II, and we arrive at the stage to prove the fact that if $\Delta$ is a connected component of the counter-image of $G$: $|w - w_0| < \varrho$, whose boundary consists of some simple curves (boundary relative to $D$) and a compact subset $e$ of $E$, then the Riemannian image $\Phi_\Delta$ of $\Delta$ covers every point of $G$ except for a (relatively) closed set of capacity zero. Since we cannot use the theory of functions of class $(U)$ in Seidel's sense in the case of $KPA$ functions[1], we prove this fact in the following way. Assume, contrary to the assertion, that $\Phi_\Delta$ does not cover any point of a totally disconnected compact set $e_w$ of positive capacity in $G$: $|w - w_0| < \varrho$. We now denote by $\Delta(\lambda)$ the part of $\Delta$ contained in the exterior of the level curve $\Gamma_\lambda$ (in the proof of (ii)). We note that $\Phi_\Delta$ is regularly exhaustible in Ahlfors' sense: $\varliminf_{\lambda\to\infty} L(\lambda)/A(\lambda) = 0$, even when $\Phi_\Delta$ is of finite area with respect to the metric (5)[2]. On the other hand, $A(\lambda) \leqq hL(\lambda)$ in the present case for almost all $\lambda$ ($\lambda_0 \leqq \lambda < \infty$, $\lambda_0$ being a fixed positive number) by the preceding lemma. Contradiction.

Remark. Using an idea of OHTSUKA [10], we give an alternative proof for (iii). Assume that $\Phi_\Delta$ does not cover any point of a totally disconnected compact set $e_w$ of positive capacity in $G$: $|w - w_0| < \varrho$. We consider the function

$$\zeta(w) = \exp\left\{-\frac{2\pi}{d}(\omega + i\tilde{\omega})\right\}. \tag{4}$$

Then, the inverse function $w = w(\zeta)$ defines a single-valued univalent analytic function $w = \psi(\zeta)$ in the annulus $A_\zeta$: $e^{-\frac{2\pi}{d}} < |\zeta| < 1$ (cf. footnote 3, p. 115) with radial slits whose arguments $\vartheta$ in the interval $(0, 2\pi)$ form a set of measure zero. Let $G'$ be the image of $A_\zeta$ in $G$ by $w = \psi(\zeta)$ and consider the part $\Phi'_\Delta$ of $\Phi_\Delta$ above $G'$. We denote by $\Delta'$ the counter-image of $\Phi'_\Delta$ by $w = f(z)$. We can now consider the $KPA$ function $\zeta = \zeta(z)$

[1] The proof of Theorem 5, § 4, II is essentially based upon the theory of functions of class $(U)$ in Seidel's sense. It is very important to remark that this method is not available in the case of $KPA$ functions as we shall see in the next paragraph.

[2] Cf. NOSHIRO [9].

$= \psi^{-1}(f(z))$ in the open set $\Delta'$. The Riemannian image $\Psi_{\Delta'}$ of $\Delta'$ by $\zeta = \zeta(z)$ consists of at most countably many covering surfaces of the basic surface $A_\zeta$. We may assume, without loss of generality, that a quadrilateral $Q_\varepsilon$: $1 - \varepsilon < |\zeta| < 1$, $\vartheta_1 < \arg\zeta < \vartheta_2$ $(\varepsilon > 0)$ is contained in $\Psi_{\Delta'}$. Then, $\Psi_{\Delta'}$ contains a quadrilateral $Q$: $e^{-\frac{2\pi}{d}} < |\zeta| < 1$, $\vartheta_1 < \arg\zeta < \vartheta_2$ with radial slits whose arguments form a set of measure zero. Consequently, $Q$ contains the radial segment: $\zeta = |\zeta|\, e^{i\vartheta}$, $e^{-\frac{2\pi}{d}} < |\zeta| < 1$ for almost every $\vartheta$ in the interval $(\vartheta_1, \vartheta_2)$ by an extension of Gross' star theorem[1]. Since the image of the segment by the $KQC$ mapping $z = z(\zeta)$ of $Q$ is an asymptotic path in $\Delta'$ starting from a point of the relative boundary of $\Delta$ and terminating at some point of $e$ of capacity zero, this is a contradiction.

6. Let $\zeta = T(z)$ be a $KQC$ mapping of $|z| < 1$ onto $|\zeta| < 1$. It had been an open question whether the extended $T$ (cf. Theorem 3) is necessarily absolutely continuous on $|z| = 1$. This question has been settled by Beurling and Ahlfors [1]:

The answer is in the *negative*. More precisely, Beurling and Ahlfors have proved that *for each $K > 1$, there exists a $KQC$ mapping of $|z| < 1$ onto $|\zeta| < 1$ such that a certain set $E_z$ of linear measure zero on $\Gamma_z$: $|z| = 1$ is mapped onto a set $E_\zeta$ of linear measure $2\pi$ on $\Gamma_\zeta$: $|\zeta| = 1$.*[2] This is a striking result, from the view-point of the theory of cluster sets. As is pointed out by Jenkins [3], we cannot extend the classical Fatou-Riesz theorem on bounded analytic functions to the case of $KPA$ functions, without any additional condition[3]. We construct an example, applying the result of Beurling and Ahlfors. Let $\zeta = T(z)$ be a $KQC$ mapping $(K > 1)$ of $|z| < 1$ onto $|\zeta| < 1$ such that a certain compact set $E_z$ of measure zero on $\Gamma_z$: $|z| = 1$ is mapped onto a compact set $E_w$ of positive measure on $\Gamma_\zeta$: $|\zeta| = 1$. We may assume that $E_z$ and $E_\zeta$ are totally disconnected perfect sets. First, we construct the harmonic measure $u(\zeta)$ of $E_\zeta$ with respect to the unit disc $|\zeta| < 1$. Then, $u(\zeta)$ is continuous and equal to 0 on the (relatively open) set $\Gamma_\zeta - E_\zeta$. Further, the angular limit $u(e^{i\vartheta})$ is equal to 1 at almost every point $e^{i\vartheta} \in E_\zeta$. Let $E'_\zeta$ denote the set of all the points $e^{i\vartheta} \in E_\zeta$ for which $u(e^{i\vartheta}) = 1$. Without loss of generality, we may assume that $E'_\zeta$ is everywhere dense on $E_\zeta$[4]. We consider the

[1] Cf. Kaplan [1, 2].

[2] Compared with Theorem 5, we see that the set $E_z$ is of positive capacity.

[3] Cf. Shibata [2].

[4] Let the closure of $E'_\zeta$ be denoted by $E^*_\zeta$. Then, obviously $E^*_\zeta \subset E_\zeta$ and $E_\zeta - E^*_\zeta$ is of measure zero. Hence, $u(\zeta)$ is identical with the harmonic measure of $E^*_\zeta$ with respect to $|\zeta| < 1$. In other words, $u(\zeta)$ can be continued to be harmonic on $\Gamma_\zeta - E^*_\zeta$ where $u(\zeta) = 0$; hence, $E^*_\zeta$ is also a perfect set. Without loss of generality, we may assume that $E_\zeta - E^*_\zeta$ is empty.

function $w = \varphi(\zeta) = 1/\exp[u(\zeta) + iv(\zeta)]$ in $|\zeta| < 1$, where $v(\zeta)$ is its conjugate. Then, by the monodromy theorem, $w = \varphi(\zeta)$ is single-valued, analytic and bounded: $e^{-1} < |\varphi(\zeta)| < 1$ in $|\zeta| < 1$. Evidently, $w = \varphi(\zeta)$ is continuous and $|\varphi(\zeta)| = 1$ on the (open) set $\Gamma_\zeta - E_\zeta$ and for almost every $e^{i\theta} \in E_\zeta$, $|\varphi(e^{i\theta})| = e^{-1}$; and, the set of singularities on $|\zeta| = 1$ is identical with $E_\zeta$. We now consider the composed function $w = f(z) \equiv \varphi[T(z)]$ in the disc $|z| < 1$. Then $w = f(z)$ is a $KPA$ function which satisfies: (i) $f(z)$ is continuous and $|f(z)| = 1$ on $\Gamma_z - E_z$; (ii) $f(z)$ is discontinuous at every point $e^{i\theta}$ of $E_z$, since the counter-image $T^{-1}(E'_\zeta) = E'_z$ is everywhere dense on $E_z$; (iii) for every $e^{i\theta} \in E'_z$ the angular limit of $|f(z)|$ is equal to $e^{-1}$ (cf. Theorem 4) [1].

Remark. The function $w = f(z) \equiv \varphi(T(z))$ is regarded as a $K$-pseudo-analytic function of class $(U)$ in Seidel's sense, but the set of values of $w = f(z)$ does not cover any point of the disc $|w| < 1/e$. Accordingly, we cannot extend the theory of functions of class $(U)$ to the case of pseudo-analytic functions without any additional condition [2].

Let $E$ be a totally disconnected compact set in a domain $G$ and let $w = T(z)$ be a $KQC$ mapping of the domain $G - E$. STREBEL [1] has proved: *In order that $w = T(z)$ can be extended to be a $KQC$ mapping of $G$, it is necessary and sufficient that the complementary domain of $E$ with respect to the $z$-plane belongs to $O_{AD}$.* Hence, *for plane domains*, the class $O_{AD}$ is preserved under quasiconformal mapping. On the contrary, the Beurling-Ahlfors result implies that for plane domains, $O_{AB}$ is not preserved under quasiconformal mapping. We have already stated that the class $O_G$ is preserved under quasiconformal mapping. However, *for open Riemann surfaces, the class $O_{HD}$ is also preserved* (ROYDEN [3], NAKAI [1]). As far as the writer knows, *whether or not $O_{HB}$ is preserved under quasiconformal mapping is still open.* It is known that, for open Riemann surfaces, the classes $O_{AB}$ and $O_{AD}$ are not preserved under quasiconformal mappings (MORI [3], ROYDEN [4]).

Although almost all results related to a compact set of capacity zero in II can be extended to the case of $KPA$ functions. On the contrary, it is impossible to extend the results based on the theory of functions of class $(U)$, the theory of meromorphic functions of bounded type and Lusin-Privaloff's theorem in III to the case of $KPA$ functions. However, it is easy to see that we can extend the theory of normal meromorphic functions introduced by LEHTO and VIRTANEN [1] to the case

[1] By the reflection principle, $w = f(z)$ can be extended to be a $KPA$ function in the complement $D_z$ of $E_z$ with respect to the $z$-plane and the Riemannian image of $D_z$ by the extended $KPA$ function $w = f(z)$ is a covering surface $\Phi_w$ of the annulus $e^{-1} < |w| < e$. We note that $w = f(z)$ is discontinuous at every point $e^{i\theta}$ of $E_z$. This example is of some interest in contrast with L. Myrberg's example [4].

[2] Cf. STORVICK [2], LOHWATER [7, 9].

of *KPA* functions by Mori's theorems [9] on normal families of *KPA* functions and Theorem 4.

# Bibliography

AGMONS, S.: [1] A property of quasi-conformal mappings. J. Rat. Mech. Analy. 3, 763—765 (1954).

AHLFORS, L. V.: [1] Sur les domaines dans lesquels une fonction méromorphe prend des valeurs appartenant à une region donnée. Acta Soc. Sci. Fenn. Nova Ser. A. 2. No. 2, 1—17 (1933). — [2] Zur Theorie der Überlagerungsflächen. Acta Math. 65, 157—194 (1935). — [3] Bounded analytic functions. Duke Math. J. 14, 1—11 (1947). — [4] Open Riemann surfaces and extremal problems on compact subregions. Comment. Math. Helvet. 24, 100—134 (1950). — [5] Remarks on the classification of open Riemann surfaces. Ann. Acad. Sci. Fenn. A. I. 87, 1—8 (1951). — [6] On the characterization of hyperbolic Riemann surfaces. Ann. Acad. Sci. Fenn. A. I. 125, 1—5 (1952). — [7] Complex analysis. McGraw-Hill-Book Company, Inc. (1953). — [8] On quasiconformal mappings. J. d'analyse Math. 3, 1—58 (1953—1954); Correction to "on quasiconformal mappings". J. d'analyse Math. 3, 207—208 (1953—1954). — [9] Conformality with respect to Riemannian metrics. Ann. Acad. Sci. Fenn. A. I. 206, 1—22 (1955).

— and A. BEURLING: [1] Conformal invariants and function-theoretic null-sets. Acta Math. 82, 101—129 (1950).

— and H. L. ROYDEN: [1] A counterexample in the classification of open Riemann surfaces. Ann. Acad. Sci. Fenn. A. I. 120 ,1—5 (1952).

BADER, R., et M. PARREAU: [1] Domaines non compacts et classification des surfaces de Riemann. C. R. Acad. Sci. (Paris) 232, 138—139 (1951).

BAGEMIHL, F.: [1] Curvilinear cluster sets of arbitrary functions. Proc. nat. Acad. Sci. (Wash.) 41, 379—382 (1955). — [2] On power series with unbounded cluster sets, and functions of class $H_2$ with meager sets of radial continuity. Proc. nat. Acad. Sci. (Wash.) 42, 763—765 (1956). — [3] On the set of values assumed by holomorphic functions near essential singularities. Math. Z. 67, 49—50 (1957). — [4] A note on power series and area. Mich. Math. J. 3, 133—135 (1955—1956). — [5] On power series, area, and length. Mich. Math. J. 4, 281—283 (1957).

— P. ERDÖS and W. SEIDEL: [1] Sur quelques propriétés frontières des fonctions holomorphes définies par certain produits dans le cercle-unité. Ann. École Norm. Sup. 70, 135—147 (1953).

— and W. SEIDEL: [1] A general principle involving Baire category, with applications to function theory and other fields. Proc. nat. Acad. Sci. (Wash.) 39, 1068—1075 (1953). — [2] Spiral and other asymptotic paths, and paths of complete indetermination, of analytic and meromorphic functions. Proc. nat. Acad. Sci. (Wash.) 39, 1251—1258 (1953). — [3] Some boundary properties of analytic functions. Math. Z. 61, 186—199 (1954). — [4] A problem concerning cluster sets of analytic functions. Math. Z. 62, 99 —110 (1955). — [5] Regular functions with prescreibed measurable boundary values almost everywhere. Proc. nat. Acad. Sci. (Wash.) 41, 740—742 (1955). — [6] Functions of bounded characteristic with prescribed ambiguous points. Mich. Math. J. 3, 77—81 (1955—1956). — [7] Values évités, mais non asymptotiques, pour les fonctions holomorphes dans le cercle-unité. C. R. Acad. Sci. (Paris) 241, 1195—1196 (1955). — [8] Some remarks on boundary behaviour of analytic and meromorphic functions. Nagoya Math. J. 9, 79—85 (1955).

BERS, L.: [1] On a theorem of Mori and the definition of quasiconformality. Trans. Amer. math. Soc. **84**, 78—84 (1957).

— and L. NIRENBERG: [1] On a representation theorem for linear elliptic systems with discontinuous coefficients and its applications. Convegno Internazionale sulle Equazioni Derivate e Parziali, Agosto 1954, pp. 111—140.

BESICOVITCH, A. S.: [1] On sufficient conditions for a function to be analytic, and on the behaviour of analytic functions in the neighbourhood of non-isolated singular points. Proc. London Math. Soc. (2) **32**, 1—9 (1931).

BEURLING, A.: [1] Étude sur un problème de majoration. Thèse, Upsal (1933). — [2] Ensembles exceptionnels. Acta Math. **72**, 1—13 (1940).

— and L. V. AHLFORS: [1] The boundary correspondence under quasiconformal mappings. Acta Math. **96**, 125—142 (1956).

BRELOT, M.: [1] Sur l'allure à la frontière des fonctions harmoniques, sous-harmoniques ou holomorphes. Bull. Soc. roy. Sci. **8**, Liège, 468—477 (1939). — [2] Familles de Perron et problème de Dirichlet. Acta Szeged. **9**, 133—153 (1939). — [3] La théorie moderne du potentiel. Ann. Inst. Fourier **4**, 113—140 (1952). — [4] On the behaviour of harmonic functions in the neighborhood of an irregular boundary point. J. d'Analyse Math. **4**, 209—221 (1954—1956). — [5] Le problème de Dirichlet. Axiomatique et frontière de Martin. J. Math. pures appl. **35**, 297—335 (1956). — [6] Axiomatique du problème de Dirichlet dans les espaces localement compacts. Séminaire de Théorie du Potentiel, 1957. Inst. de H. Poincaré. — [7] Sur l'allure à la frontière des fonctions sous-harmoniques ou holomorphes. Ann. Acad. Sci. Fenn. A. I. **250**/4, 1—9 (1958).

— et G. CHOQUET: [1] Espaces et lignes de Green. Ann. Inst. Fourier **3**, 199—263 (1951).

CACCIOPPOLI, R.: [1] Funzioni pseudo-analitiche e rapprentazioni pseudo-conformi delle superficie riemanniane. Recerche Mat. **2**, 104—127 (1953).

CALDÉRON, A. P., A. GONZÁLEZ-DOMÍNGUEZ and A. ZYGMUND: [1] Nota sobre les valores limites de funciones analiticas. Rev. Union Mat. Argent. **14**, 16—19 (1949).

CARATHÉODORY, C.: [1] Conformal representation. Cambridge, 1932. — [2] Zum Schwarzschen Spiegelungsprinzip (die Randwerte von meromorphen Funktionen). Comment. Math. Helvet. **19**, 263—278 (1946—1947).

CARTWRIGHT, M. L.: [1] On the behaviour of an analytic function in the neighbourhood of its essential singularities. Math. Ann. **112**, 161—187 (1936). — [2] On the asymptotic values of functions with a non-enumerable set of essential singularities. J. London Math. Soc. **11**, 303—306 (1936). — [3] The exceptional values of functions with a non-enumerable set of essential singularities. Quart. J. Math. **8**, 303—307 (1937).

COLLINGWOOD, E. F.: [1] Exceptional values of meromorphic functions. Trans. Amer. math. Soc. **66**, 308—346 (1949). — [2] Sur les ensembles d'accumulation radiaux et angulaires des fonctions analytiques. C. R. Acad. Sci. (Paris) **238**, 1769—1771 (1954). — [3] On the linear and angular cluster sets of functions meromorphic in the unit circle. Acta Math. **91**, 165—185 (1954). — [4] Sur le comportement à la frontière d'une fonction méromorphe dans le cercle unité. C. R. Acad. Sci. (Paris) **240**, 1502—1504 (1955). — [5] Sur les ensembles d'indétermination maximum des fonctions analytiques. C. R. Acad. Sci. (Paris) **240**, 1604—1606 (1955). — [6] A theorem on certain classes of singularities defined by cluster sets. J. London Math. Soc. **30**, 422—424 (1955). — [7] On a theorem of Eggleston concerning cluster sets. J. London Math. Soc. **30**, 425—428 (1955). — [8] A theorem on prime ends. J. London Math. Soc. **31**, 344—349 (1956). — [9] On sets of maximum indetermination of analytic functions. Math. Z. **67**, 377—396 (1957). — [10] Addendum: On sets of maximum indetermination of analytic functions. Math. Z. **68**, 498—499 (1958). — [11] Cluster sets and prime ends. Ann. Acad. Sci. Fenn. A. I. **250/6**, 1—11 (1958).

— and M. L. Cartwright: [1] Boundary theorems for a function meromorphic in the unit circle. Acta Math. **87**, 83—146 (1952).

— and A. J. Lohwater: [1] Inégalités relatives aux défauts d'une fonction méromorphe dans le cercle unité. C. R. Acad. Sci. (Paris) **242**, 1255—1257 (1956). — [2] Applications of the theory of cluster sets to a class of meromorphic functions. Proc. Cambridge Phil. Soc. **53**, Part 1, 93—105 (1957).

Constantinescu, C.: [1] Einige Anwendungen des hyperbolischen Maßes. Math. Nachr. **15**, 155—172 (1956).

— u. A. Cornea: Über den idealen Rand und einige seiner Anwendungen bei der Klassifikation der Riemannschen Flächen. Nagoya Math. J. **13**, 169—233 (1958).

Cornea, A.: [1] On the behaviour of analytic functions in the neighborhood of the boundary of a Riemann surface. Nagoya Math. J. **12**, 55—58 (1957). — [2] Über eine Formel in der Extremisierungstheorie. Rev. Math. pur. appl. **3**, 431—436 (1958).

Doob, J. L.: [1] On a theorem of Gross and Iversen. Ann. of Math. **33**, 753—757 (1932). — [2] The boundary values of analytic functions. Trans. Amer. math. Soc. **34**, 153—170 (1932). — [3] The boundary values of analytic functions. Trans. Amer. math. Soc. **35**, 418—451 (1933). — [4] The ranges of analytic functions. Ann. of Math. **36**, 117—126 (1935). — [5] Semi-martingales and subharmonic functions. Trans. Amer. math. Soc. **77**, 86—121 (1954). — [6] Probability methods applied to the first boundary value problem. Proc. Third Berkeley Symp. on Math. Statist. and Probab. **2**, 49—80 (1954—1955).

Dufresnoy, J.: [1] Sur les domaines couverts par les valeurs d'une fonction méromorphe ou algebroïde. Ann. École Norm. Sup. (3) **58**, 179—259 (1941).

Eggleston, H. G.: [1] The range set of a function meromorphic in the unit circle' Proc. London Math. Soc. (3) **5**, 500—512 (1955). — [2] A property of bounded analytic functions. Comment. Math. Helvet. **30**, 139—143 (1956).

Evans, G. C.: [1] Potentials and positively infinite singularities of harmonic functions. Mh. Math. u. Phys. **43**, 419—424 (1936).

Fatou, P.: [1] Séries trigonométriques et séries de Taylor. Acta Math. **30**,335—400 (1906).

Fekete, M.: [1] Über die Verteilung der Wurzeln bei gewissen algebraischen Gleichungen mit ganzzahligen Koeffizienten. Math. Z. **17**, 228—249 (1923).

Frostman, O.: [1] Potentiel d'équilibre et capacité des ensembles. Thèse, Lund, 1935. — [2] Sur les produits de Blaschke. Fysiogr. Sällsk. I Lund Förh. **12**, Nr. 15, 169—182 (1942).

Gross, W.: [1] Über die Singularitäten analytischer Funktionen. Mh. Math. u. Phys. **29**, 3—47 (1918). — [2] Zum Verhalten analytischer Funktionen in der Umgebung singulärer Stellen. Math. Z. **2**, 243—294 (1918). — [3] Einige ganze Funktionen, für die jede komplexe Zahl Konvergenzwert ist. Math. Ann. **79**, 201—208 (1918).

Grötzsch, H.: [1] Über die Verzerrung bei schlichten nichtkonformen Abbildungen und über eine damit zusammenhängende Erweiterung des Picardschen Satzes. Leipziger Ber. **80**, 503—507 (1928). — [2] Über die Verzerrung bei nichtkonformen schlichten Abbildungen mehrfach zusammenhängender schlichter Bereiche.

Leipziger Ber. **82**, 69—90 (1930). — [3] Über möglichst konforme Abbildungen von schlichten Bereichen. Leipziger Ber. **84**, 114—120 (1932).

Hällström, G. af: [1] Über eindeutige analytische Funktionen mit unendlich vielen wesentlichen Singularitäten. 9. Congr. Math. Scand. Helsingfors, 1938. — [2] Über meromorphe Funktionen mit mehrfach zusammenhängenden Existenzgebieten. Acta Acad. Abo. Math. et Phys. **12**, 5—100 (1939). — [3] Zur Berechnung der Bodenordnung oder Bodenhyperordnung eindeutiger Funktionen. Ann. Acad. Sci. Fenn. A. I. **193**, 1—16 (1955). — [4] Übertragung eines Satzkomplexes von Weierstraß und Dinghas auf beliebige Randmengen der Kapazität Null. Ann. Acad. Sci. Fenn. A. I. **250/12**, 1—9 (1958). — [5] Wertverteilungssätze pseudomeromorpher Funktionen. Acta Acad. Abo. **21**, no. 9, 1—23 (1958).

Heins, M.: [1] The conformal mapping of simply-connected Riemann surfaces. Ann. of Math. **50**, 686—690 (1949). — [2] Riemann surfaces of infinite genus. Ann. of Math. **55**, 296—317 (1952). — [3] Studies in the conformal mapping of Riemann surfaces. I. Proc. nat. Acad. Sci. (Wash.) **39**, 322—324 (1953). — [4] The set of asymptotic values of an entire function. 12. Congr. Math. Scand. Lund. 56—60 (1953). — [5] Studies in the conformal mapping of Riemann surfaces. II. Proc. nat. Acad. Sci. (Wash.) **40**, 302—305 (1954). — [6] Meromorphic functions with assigned asymptotic values. Duke Math. J. **22**, 353—356 (1955). — [7] A universal Blaschke product. Arch. Math. **6**, 41—44 (1955). — [8] On the Lindelöf principle. Ann. of Math. **61**, 440—473 (1955). — [9] Lindelöfian maps. Ann. of Math. **62**, 418—446 (1955).

Herglotz, G.: [1] Über Potenzreihen mit positivem, reellen Teil im Einheitskreise. Leipziger Ber. **63**, 501—511 (1911).

Hersch, J.: [1] Contribution à la théorie des fonctions pseudoanalytiques. Comment. Math. Helvet. **30**, 1—19 (1956).

— et A. Pfluger: [1] Généralisation du lemme de Schwarz et du principe de la mesure harmonique pour les fonctions pseudoanalytiques. C. R. Acad. Sci. (Paris) **234**, 43—45 (1952).

Hervé, M.: [1] A propos d'un mémoire récent de M. Noshiro: Nouvelles applications de sa méthode. C. R. Acad. Sci. (Paris) **232**, 2170—2172 (1951). — [2] Sur les valeurs omises par une fonction méromorphe. C. R. Acad. Sci. (Paris) **240**, 718—720 (1955). — [3] Contribution à l'étude d'une fonction méromorphe au voisinage d'un ensemble singulier de capacité nulle. J. Math. pures appl. **35**, 161—173 (1956). — [4] Valeurs exceptionnelles d'une fonction méromorphe au voisinage d'un ensemble singulier de capacité nulle. Ann. Acad. Sci. Fenn. A. I. **250/14**, 1—4 (1958).

Herzog, F., and G. Piranian: [1] Sets of radial continuity of analytic functions. Pacific J. Math. **4**, 533—538 (1954).

Hong, I.: [1] On positively infinite singularities of a solution of the equation $\Delta u + k^2 u = 0$. Kôdai Math. Sem. Rep. **8**, 9—12 (1956).

Hössjer, G.: [1] Über die Randwerte beschränkter Funktionen. Acta Szeged **5**, 55 (1930).

— u. O. Frostman: [1] Über die Ausnahmestellen eines Blaschkeproduktes. Medd. fran Lunds Univ. Math. Sem. **1** (1933).

Huckemann, F.: [1] Über den Defekt von mittelbaren Randstellen auf beschränktartigen Riemannschen Flächen. Ann. Acad. Sci. Fenn. A. I. **250/16**, 1—12(1958).

Inoue, M.: [1] Positively infinite singularities of solutions of linear elliptic partial differential equations. J. Inst. Polytechnics. Osaka City Univ. **8**, 43—50 (1957).

Irie, S.: [1] Sur un théorème de M. Beurling. Proc. Japan Acad. **13**, 244—246 (1937).

IVERSEN, F.: [1] Recherches sur les fonctions inverses des fonctions méromorphes. Thèse, Helsingfors (1914), pp. 1—67. — [2] Sur quelque propriétés des fonctions monogènes au voisinage d'un point singulier. Öfersigt af Finska Vet.-Soc. Förh. **58 A**, No. 25 (1915—1916). — [3] Zum Verhalten analytischer Funktionen in Bereichen, deren Rand eine wesentliche Singularität enthält. Öfersigt af Finska Vet.-Soc. Förh. **44 A**, Nr. 4 (1921).

JENKINS, J. A.: [1] On a problem of Lusin. Mich. Math. J. **3**, 187—189 (1955—1956). — [2] On quasiconformal mappings. J. Rat. Mech. Analy. **5**, 343—352 (1956). — [3] A new criterion for quasiconformal mapping. Ann. of Math. **65**, 208—214 (1957).

JUVE, Y.: [1] Über gewisse Verzerrungseigenschaften konformer und quasikonformer Abbildungen. Ann. Acad. Sci. Fenn. A. I. **174**, 1—40 (1954).

KAKUTANI, S.: [1] Applications of the theory of pseudo-regular functions to the type-problem of Riemann surfaces. Jap. J. Math. **13**, 375—392 (1936). — [2] Random walk and the type problem of Riemann surfaces. Contributions to the Theory of Riemann surfaces. Ann. Math. Studies 30. Princeton: Princeton University Press 1953.

KAMETANI, S.: [1] The exceptional values of functions with the set of linear measure zero of essential singularities. Proc. Japan Acad. **17**, 117—120 (1941). — [2] The exceptional values of functions with the set of capacity zero of essential singularities. Proc. Japan Acad. **17**, 429—433 (1941). — [3] The exceptional values of functions with the set of linear measure zero of essential singularities II. Proc. Japan Acad. **19**, 438—443 (1943). — [4] On Hausdorff's measures and generalized capacities with some of their applications to the theory of functions. Jap. J. Math. **19**, 217—257 (1945).

— and T. UGAERI: [1] A remark on Kawakami's extension of Löwner's lemma. Proc. Japan Acad. **18**, 14—15 (1942).

KAPLAN, W.: [1] On Gross' star theorem, schlicht functions, logarithmic potentials and Fourier series. Ann. Acad. Sci. Fenn. A. I. **86**, 1—23 (1951). — [2] Extensions of the Gross star theorem. Mich. Math. J. **2**, 105—108 (1953—1954).

KAWAKAMI, Y.: [1] On an extension of Löwner's lemma. Jap. J. Math. **17**, 569—572 (1941).

KIERST, S.: [1] Sur l'ensemble des valeurs asymptotiques d'une fonction méromorphe dans le cercle-unité. Fund. Math. **27**, 226—233 (1936).

— et E. SZPILRAJN: [1] Sur certaines singularités des fonctions analytiques uniformes. Fund. Math. **21**, 276—294 (1933).

KOEBE, P.: [1] Randerzuordnung bei konformer Abbildung. Göttinger Nachr. 286—288 (1913). — [2] Abhandlungen zur Theorie der konformen Abbildung I. J. reine angew. Math. **146**, 177—225 (1915).

KUNUGUI, K.: [1] Sur un théorème de MM. Seidel-Beurling. Proc. Japan Acad. **15**, 27—32 (1939). [2] Sur un problème de M. A. Beurling. Proc. Japan Acad. **16**, 361—366 (1940). — [3] Sur l'allure d'une fonction analytique uniforme au voisinage d'un point frontière de son domaine de definition. Jap. J. Math. **18**, 1—39 (1942).

KÜNZI, H.: [1] Quasikonforme Abbildungen. Ann. Acad. Sci. Fenn. A. I. **249**/2, 1—24 (1958).

KURAMOCHI, Z.: [1] On covering surfaces. Osaka Math. J. **5**, 155—201 (1953). — [2] Relations between harmonic dimensions. Proc. Japan Acad. **30**, 576—580 (1954). — [3] On the behaviour of analytic functions on abstract Riemann surfaces. Osaka Math. J. **7**, 109—127 (1955). — [4] On the ideal boundaries of abstract Riemann surfaces. Osaka Math. J. **10**, 83—102 (1958). — [5] Cluster sets of analytic functions in open Riemann surfaces with regular metrics I. Osaka Math. J. 11, 83—90 (1959).

— and T. KURODA: [1] A note on the set of logarithmic capacity zero. Proc. Japan Acad. **30**, 566—569 (1954).

KURATOWSKI, C.: [1] Topologie I. 2. ed. Warszawa-Wrocław (1948). — [2] Topologie II (1950).

KURODA, T.: [1] On the type of an open Riemann surface. Proc. Japan Acad. **27**, 57—60 (1951). — [2] Some remarks on an open Riemann surface with null boundary. Tôhoku Math. J. **3**, 182—186 (1951). — [3] On the uniform meromorphic functions with the set of capacity zero of essential singularities. Tôhoku Math. J. **3**, 257—269 (1951). — [4] A property of some open Riemann surfaces and its application. Nagoya Math. J. **6**, 77—84 (1953). — [5] On the classification of symmetric Fuchsian groups of genus zero. Proc. Japan Acad. **29**, 431—434 (1953). — [6] Theorems of the Phragmén-Lindelöf type on an open Riemann surface. Osaka Math. J. **6**, 231—241 (1954). — [7] On analytic functions on some Riemann surfaces. Nagoya Math. J. **10**, 27—50 (1956). — [8] Remarks on some covering surfaces. Rev. Math. pures appl. **2**, 239—244 (1957).

LAVRENT'EV, M.: [1] A fundamental theorem of the theory of quasi-conformal mapping of plane regions. Izvestiya Akad. Nauk SSSR. Ser. Mat. **12**, 513—554 (1948). Russian. Amer. Math. Soc. Transl. 29.

LEHTO, O.: [1] A majorant principle in the theory of functions. Math. Scand. **1**, 5—17 (1953). — [2] Sur la théorie des fonctions méromorphes à caractéristique bornée. C. R. Sci. Acad. (Paris) **236**, 1943—1945 (1953). — [3] On meromorphic functions whose values lie in a given domain. Ann. Acad. Sci. Fenn. A. I. **160**, 1—14 (1953). — [4] On an extension of the concept of deficiency in the theory of meromorphic functions. Math. Scand. **1**, 207—212 (1953). — [5] On the distribution of values of meromorphic functions of bounded characteristic. Acta Math. **91**, 87—112 (1954). — [6] On meromorphic functions of bounded characteristic. 12. Congr. Math. Scand. Lund (1953) 183—187. — [7] Value distribution and boundary behaviour of a function of bounded characteristic and the Riemann surface of its inverse function. Ann. Acad. Sci. Fenn. A. I. **177**, 1—46 (1954). — [8]On the first boundary value problem for functions harmonic in the unit circle. Ann. Acad. Sci. Fenn. I. A. **210**, 1—26 (1955). — [9] Distribution of values and singularities of analytic functions. Ann. Acad. Sci. Fenn. A. I. **249**/3, 1—14 (1957). — [10] A generalization of Picard's theorem. Arkiv. för Mat. **3**, 495—500 (1958).

— and K. I. VIRTANEN: [1] Boundary behaviour and normal meromorphic functions. Acta Math. **97**, 47—65 (1957). — [2] On the behaviour of meromorphic functions in the neighborhood of an isolated singularity. Ann. Acad. Sci.Fenn. A. I. **240**, 1—9 (1957).

LOHWATER, A. J.: [1] A uniqueness theorem for a class of harmonic functions. Proc. Amer. math. Soc. **3**, 278—279 (1952). — [2] The boundary values of a class of meromorphic functions. Duke Math. J. **19**, 243—252 (1952). — [3] Les valeurs asymptotiques de quelques fonctions méromorphes dans le cercle-unité. C. R. Acad. Sci. (Paris) **237**, 16—18 (1953). — [4] On the Schwarz reflection principle. Mich. Math. J. **2**, 151—156 (1953—1954). — [5] On the radial limits of analytic functions. Proc. Amer. math, Soc. **6**, 79—83 (1955). — [6] Sur le principe de symétrie et la répartition des valeurs des fonctions analytiques bornées. C. R. Acad. Sci. (Paris) **242**, 2278—2281 (1956). — [7] The reflection principle and the distribution of values of functions defined in a circle. Ann. Acad. Sci. Fenn. A. I.

229, 1—18 (1956). — [8] The boundary behaviour of a quasiconformal mapping. J. Rat. Mech. Analy. **5**, 335—342 (1956). — [9] The boundary behaviour of meromorphic functions. Ann. Acad. Sci. Fenn. A. I. **250/22**, 1—6 (1958).

— and G. Piranian: [1] Conformal mapping of a Jordan region whose boundary has positive two-dimensional measure. Mich. Math. J. **1**, 1—4 (1952). — [2] On the derivative of a univalent function. Proc. Amer. math. Soc. **4**, 591—594 (1953). — [3] On a conjecture of Lusin. Mich. Math. J. **3**, 63—68 (1955—1956).

— and W. Seidel: [1] An example in conformal mapping. Duke Math. J. **15**, 137—143 (1948).

Lokki, O.: [1] Über das Randwertproblem der analytischen Funktionen. Ann. Acad. Sci. Fenn. A. I. **144**, 1—8 (1952).

Lusin, N., et J. Privaloff: [1] Sur l'unicité et la multiplicité des fonctions analytiques. Ann. Sci. École Norm. Sup. (3) **42**, 143—191 (1925).

Martin, R. S.: [1] Minimal positive harmonic functions. Trans. Amer. math. Soc. **49**, 137—172 (1941).

Marty, F.: [1] Recherches sur la répartition des valeurs d'une fonction méromorphe. Ann. Fac. Sci. Univ. Toulouse (3) **23**, 183—261 (1931).

Matsumoto, K.: [1] Remarks on some Riemann surfaces. Proc. Japan Acad. **34**, 672—675 (1958). — [2] On subsurfaces of some Riemann surfaces. Nagoya Math. J. **15**, 261—274 (1959).

Meier, K.: [1] Über die Randwerte meromorpher Funktionen und hinreichende Bedingungen für Regularität von Funktionen einer komplexen Variablen. Comment. Math. Helvet. **24**, 238—259 (1950). — [2] Über Mengen von Randwerten meromorpher Funktionen. Comment. Math. Helvet. **30**, 224—233 (1955).

Mergelyan, S. N.: [1] On the representation of functions by series of polynomials on closed sets (in Russian). Doklady Akad. Nauk SSSR, N. S. **78**, 405—408 (1951). Amer. Math. Soc. Transl. No. **85**, Providence (1953). — [2] Uniform approximations to functions of a complex variable (in Russian). Uspekhi Mat. Nauk, N. S. **7**, No. 2 (48), 31—122 (1952). Amer. Math. Soc. Transl. No. 101. Providence (1954).

Mori, A.: [1] On a conformal mapping with certain boundary correspondences. J. Math. Soc. Japan **2**, 129—132 (1950). — [2] On Riemann surfaces on which no bounded harmonic function exists. J. Math. Soc. Japan **3**, 285—288 (1951). — [3] On the existence of harmonic functions on a Riemann surface. J. Fac. of Sci. Univ. of Tokyo, I **6**, 247—257 (1951). — [4] A remark on the class $O_{HD}$ of Riemann surfaces. Kôdai Math. Sem. Rep. 57—58 (1952). — [5] A remark on the prolongation of Riemann surfaces of finite genus. J. Math. Soc. Japan **4**, 27—30 (1952). — [6] Conformal rigidity of Riemann surfaces. J. Math. Soc. Japan **4**, 302—309 (1952). — [7] An imbedding theorem on finite covering surfaces. J. Math. Soc. Japan **5**, 263—268 (1953). — [8] On an absolute constant in the theory of quasi-conformal mappings. J. Math. Soc. Japan **8**, 156—166 (1956). — [9] On quasi-conformality and pseudo-analyticity. Trans. Amer. math. Soc. **84**, 56—77 (1957).

Morrey, jr., C. B.: [1] On the solutions of quasilinear elliptic partial differential equations. Trans. Amer. math. Soc. **43**, 126—166 (1938).

Myrberg, L.: [1] Bemerkungen zur Theorie der harmonischen Funktionen. Ann. Acad. Sci. Fenn. A. I. **107**, 1—8 (1952). — [2] Über das Verhalten der Greenschen Funktionen in der Nähe des idealen Randes einer Riemannschen Fläche. Ann. Acad. Sci. Fenn. A. I. **139**, 1—8 (1952). — [3] Über die Existenz von positiven harmonischen Funktionen auf Riemannschen Flächen. Ann. Acad. Sci. Fenn.

A. I. **146**, 1—6 (1953). — [4] Eine Bemerkung zum Picardschen Satz. Ann. Acad. Sci. Fenn. A. I. **255**, 1—4 (1958).

MYRBERG, P. J.: [1] Ein Approximationssatz für die Fuchsschen Gruppen. Acta Math. **57**, 389—409 (1931). — [2] Über die Existenz der Greenschen Funktionen auf einer gegebenen Riemannschen Fläche. Acta Math. **61**, 39—79 (1933). — [3] Die Kapazität der singulären Menge der linearen Gruppen. Ann. Acad. Sci. Fenn. A. I. **11**, 1—7 (1942). — [4] Über die analytische Fortsetzung von beschränkten Funktionen. Ann. Acad. Sci. Fenn. A. I. **58**, 1—7 (1949). — [5] Über die Existenz von beschränktartigen automorphen Funktionen. Ann. Acad. Sci. Fenn. A. I. **77**, 1—7 (1950). — [6] Reduktion der Verzweigungspunkte Riemannscher Flächen durch konforme Abbildung. Ann. Acad. Sci. Fenn. A. I. **261**, 1—7 (1959).

NAÏM, L.: [1] Sur le rôle de la frontière de R. S. Martin dans la théorie du potentiel. Ann. Inst. Fourier **7**, 5—103 (1957).

NAKAI, M.: [1] On a ring isomorphism induced by quasiconformal mappings. Nagoya Math. J. **14**, 201—221 (1959).

NEVANLINNA, R.: [1] Über beschränkte analytische Funktionen. Ann. Acad. Sci. Fenn. (A) **32**, No. 7 (1929). — [2] Quadratisch integrierbare Differentiale auf einer Riemannschen Mannigfaltigkeit. Ann. Acad. Sci. Fenn. A. I. **1**, 1—34 (1941). — [3] Über das Anwachsen des Dirichletintegrals einer offenen Riemannschen Fläche. Ann. Acad. Sci. Fenn. A. I. **45**, 1—9 (1948). — [4] Über Mittelwerte von Potentialfunktionen. Ann. Acad. Sci. Fenn. A. I. **57**, 1—12 (1949). — [5] Über die Existenz von beschränkten Potentialfunktionen auf Flächen von unendlichem Geschlecht. Math. Z. **52**, 599—604 (1950). — [6] Eindeutige analytische Funktionen. Zweite Auflage. Berlin-Göttingen-Heidelberg: Springer 1953. — [7] Uniformisierung. Berlin-Göttingen-Heidelberg: Springer 1953. — [8] A remark on differentiable mappings. Mich. Math. J. **3**, 53—57 (1955).

NOSHIRO, K.: [1] Some theorems on a cluster set of an analytic function. Proc. Japan Acad. **13**, 27—29 (1937). — [2] On the theory of the cluster sets of analytic functions. J. Fac. Sci. Hokkaido Univ. **6**, 217—231 (1937). — [3] Contributions to the theory of meromorphic functions in the unit circle. J. Fac. Sci. Hokkaido Univ. **7**, 149—159 (1938). — [4] On the singularities of analytic functions. Jap. J. Math. **17**, 37—96 (1940). — [5] On the singularities of analytic functions with a general domain of existence. Proc. Japan Acad. **22**, 233—237 (1946). — [6] Contributions to the theory of the singularities of analytic functions. Jap. J. Math. **19**, 299—327 (1948). — [7] Note on the cluster sets of analytic functions. J. Math. Soc. Japan **1**, 275—281 (1950). — [8] A theorem on the cluster sets of pseudo-analytic functions. Nagoya Math. J. **1**, 83—89 (1950). — [9] Open Riemann surface with null boundary. Nagoya Math. J. **3**, 73—79 (1951). — [10] On the theory of cluster sets of analytic functions (in Japanese). Sugaku **5**, 65—72 (1953). Amer. math. Soc. Transl. **8**, Ser. 2, 1—12 (1958). — [11] Cluster sets of functions meromorphic in the unit circle. Proc. nat. Acad. Sci. (Wash.) **41**, 398—401 (1955).

OHTSUKA, M.: [1] On the cluster sets of analytic functions in a Jordan domain. J. Math. Soc. Japan **2**, 1—15 (1950). — [2] Dirichlet problems on Riemann surfaces and conformal mapping. Nagoya Math. J. **3**, 91—137 (1951). — [3] On the behaviour of analytic function about an isolated boundary point. Nagoya Math. J. **4**, 103—108 (1952). — [4] Note on the harmonic measure of the accessible boundary of a covering Riemann surface. Nagoya Math. J. **5**, 35—38 (1953). — [5] Boundary components of Riemann surfaces. Nagoya Math. J. **7**, 65—83 (1954). — [6] Note on functions bounded and analytic in the unit circle. Proc.

Amer. math. Soc. **5**, 533—535 (1954). — [7] On exceptional values of a meromorphic function. Nagoya Math. J. **9**, 119—121 (1955). — [8] Théorèmes étoilés de Gross et leurs applications. Ann. Inst. Fourier **5**, 1—28 (1955). — [9] On asymptotic values of functions analytic in a circle. Trans. Amer. math. Soc. **78**, 294—304 (1955). — [10] Sur les ensembles d'accumulation relatif à des transformations plus générales que les transformations quasi conformes. Ann. Inst. Fourier **5**, 29—37 (1955). — [11] Sur un théorème étoilé de Gross. Nagoya Math. J. **9**, 191—207 (1955). — [12] Generalizations of Montel-Lindelöf's theorem on asymptotic values. Nagoya Math. J. **10**, 129—163 (1956). — [13] On boundary values of an analytic transformation of a circle into a Riemann surface. Nagoya Math. J. **10**, 171—175 (1956). — [14] Sur les ensembles d'accumulation relatifs à des transformations localement pseudo-analytiques au sens de Pfluger-Ahlfors. Nagoya Math. J. **11**, 131—144 (1957). — [15] On boundary cluster sets of functions analytic in the unit circle. Rev. Math. pures appl. **2**, 317—321 (1957).

Painlevé, P.: [1] Sur les singularités des fonctions analytiques et en particulier des fonctions définies par les équations différentielles. C. R. Acad. Sci. (Paris) **131**, 487—492 (1900).

Parreau, M.: [1] Sur les moyennes des fonctions harmoniques et analytiques. Thèse, Paris 1952, et Ann. Inst. Fourier **3**, 103—197 (1951). — [2] Fonction caractéristique d'une application conforme. Ann. Fac. Sci. Univ. Toulouse, 4e sér., **19**, 175—190 (1955). — [3] Théorème de Fatou et problème de Dirichlet pour les ligne de Green de certaines surfaces de Riemann. Ann. Acad. Sci. Fenn. A. I. **250/25**, 1—8 (1958).

Pfluger, A.: [1] Sur une propriété de l'application quasi-conforme d'une surface de Riemann ouverte. C. R. Acad. Sci. (Paris) **227**, 25—26 (1948). — [2] Über das Anwachsen eindeutiger analytischer Funktionen auf offenen Riemannschen Flächen. Ann. Acad. Sci. Fenn. A. I. **64**, 1—18 (1949). — [3] Quasikonforme Abbildungen und logarithmische Kapazität. Ann. Inst. Fourier **2**, 69—80 (1951). — [4] Extremallängen und Kapazität. Comment. Math. Helvet. **29**, 120—131 (1955). — [5] Über die Äquivalenz der geometrischen und der analytischen Definition quasikonformer Abbildungen. Comment. Math. Helvet. **33**, 23—33 (1959).

Phragmén, E., et E. Lindelöf: [1] Sur une extension d'un principe classique de l'analyse et sur quelques propriétés des fonctions monogènes dans le voisinage d'un point singulier. Acta Math. **31**, 381—406 (1908).

Piranian, G.: [1] Construction of functions with prescribed boundary behavior. Ann. Acad. Sci. Fenn. A. I. **250/26**, 1—8 (1958).

— and W. Rudin: [1] Lusin's theorem on areas of conformal maps. Mich. Math. J. **3**, 191—199 (1955—1956).

— and A. Shields: [1] The set of Lusin points of analytic functions. Mich. Math. J. **4**, 15—22 (1957).

Plessner, A.: [1] Über das Verhalten analytischer Funktionen am Rande ihres Definitionsbereiches. J. reine angew. Math. **158**, 219—227 (1927).

Riesz, F.: [1] Über die Randwerte einer analytischen Funktion. Math. Z. **18**, 87—95 (1923).

— u. M. Riesz: [1] Über die Randwerte einer analytischen Funktion. 4. Congr. Math. Scand. Stockholm (1916) pp. 27—47.

Roth, A.: [1] Approximationseigenschaften und Strahlengrenzwerte meromorpher und ganzer Funktionen. Comment. Math. Helvet. **11**, 77—125 (1938).

ROYDEN, H. L.: [1] Some remarks on open Riemann surface. Ann. Acad. Sci. Fenn. A. I. **85**, 1—8 (1951). — [2] Harmonic functions on open Riemann surfaces. Trans. Amer. math. Soc. **73**, 40—94 (1952). — [3] A property of quasi-conformal mapping. Proc. Amer. math. Soc. **5**, 266—269 (1954). — [4] Open Riemann surfaces. Ann. Acad. Sci. Fenn. A. I. **249/5**, 1—13 (1958).

RUDIN, W.: [1] Positive infinities of potentials. Proc. Amer. math. Soc. **2**, 967—969 (1951). — [2] On a problem of Collingwood and Cartwright. J. London Math. Soc. **30**, 232—238 (1955). — [3] Some theorems on bounded analytic functions. Trans. Amer. math. Soc. **78**, 333—342 (1955). — [4] The radial variation of analytic functions. Duke Math. J. **22**, 235—242 (1955). — [5] Boundary values of continuous analytic functions. Proc. Amer. math. Soc. **7**, 808—811 (1956).

SAKAI, E.: [1] Note on pseudo-analytic functions. Proc. Japan Acad. **25**, 12—17 (1949).

SAKS, S.: [1] Theory of the integral. 2. ed., Warsaw, 1937.

SARIO, L.: [1] Über Riemannsche Flächen mit hebbarem Rand. Ann. Acad. Sci. Fenn. A. I. **50**, 1—79 (1948). — [2] Sur la classification des surfaces de Riemann. 11. Congr. Math. Scand. Trondheim (1949) pp. 229—238. — [3] Sur le problème du type des surfaces de Riemann. C. R. Acad. Sci. (Paris) **229**, 1109—1111 (1949). — [4] Quelques proprietés à la frontière se rattachant à la classification des surfaces de Riemann. C. R. Acad. Sci. (Paris) **230**, 42—44 (1950). — [5] Questions d'existence au voisinage de la frontière d'une surface de Riemann. C. R. Acad. Sci. (Paris) **230**, 269—271 (1950). — [6] A linear operator method on arbitrary Riemann surfaces. Trans. Amer. math. Soc. **72**, 281—295 (1952). — [7] An extremal method on arbitrary Riemann surfaces. Trans. Amer. math. Soc. **72**, 459—470 (1952). — [8] Minimizing operators on subregions. Proc. Amer. math. Soc. **4**, 350—355 (1953). — [9] Modular criteria on Riemann surfaces. Duke Math. J. **20**, 141—338 (1953). — [10] Alternating method on arbitrary Riemann surfaces. Pacific J. Math. **3**, 631—645 (1953). — [11] Capacity of the boundary and of a boundary component. Ann. of Math. **50**, 135—144 (1954). — [12] Strong and weak boundary components. J. d'analyse Math. **5**, 389—398 (1956—1957).

SAVAGE, N.: [1] Weak boundary component of an open Riemann surface. Duke Math. J. **24**, 79—96 (1957).

SCHLESINGER, L., u. A. PLESSNER: [1] Lebesguesche Integrale und Fouriersche Reihen. Berlin u. Leipzig, 1926.

SEIDEL, W.: [1] On the cluster values of analytic functions. Trans. Amer. math. Soc. **34**, 1—21 (1932). — [2] On the distribution of values of bounded analytic functions. Trans. Amer. math. Soc. **36**, 201—226 (1934). — [3] Holomorphic functions with spiral asymptotic paths. Nagoya Math. J. **14**, 159—171 (1959).

SELBERG, H.: [1] Über die ebenen Punktmengen von der Kapazität Null. Avh. Norske Videnskaps-Akad. Oslo I Math.-Natur. (1937), No. 10.

SHIBATA, K.: [1] Remarks on the sequences of quasi-conformal mappings. Proc. Japan Acad. **32**, 665—670 (1956). — [2] On boundary values of some pseudo-analytic functions. Proc. Japan Acad. **33**, 628—632 (1957).

STOÏLOW, S.: [1] Sur les fonctions analytiques dont les surfaces de Riemann ont des frontières totalement discontinues. Mathematica (Cluj) **12**, 123—138 (1936). — [2] Remarques sur les fonctions analytiques continues dans un domaine ou elles admettent un ensemble parfait dinscontinu de singularités. Bull. Math. Soc. Roumaine des Sci. **38**, 117—120 (1936). — [3] Sur les surfaces de Riemann normalement exhaustibles et sur le théorème des disques pour ces surfaces. Compositio Math. 7, 428—435 (1940). — [4] Sur une extension topologique du principe du maximum du module et ses applications à la théorie des fonctions. Bull. Sect. Sci. Acad. Roumaine 23, 1—3 (1940). — [5] Sur les singularités des fonctions analytiques uniformes dont la surface de Riemann a sa frontière de mesure harmoniques nulle. Mathematica (Timisoara) 19, 126—138 (1943). — [6] Remarques sur la définition des points singuliers des fonctions analytiques multiformes. Bull. Sect. Sci. Acad. Roumaine 26, 1—2 (1944). — [7] Quelques remarques sur les éléments frontière des surfaces de Riemann et sur les fonctions correspondant à ces surfaces. C. R. Acad. Sci. (Paris) 227, 1326—1328 (1948). — [8] Note sur les fonctions analytiques multiformes. Ann. Soc. Pol. Math. 25, 69—74 (1952). — [9] Leçons sur les principes topologiques de la théorie des fonctions analytiques. 2. éd. Paris (1956). — [10] Sur la classification topologique de recouvrement riemanniens. Rev. Math. pures appl. 1, 37—42 (1956). — [11] Sur la théorie topologique des recouvrements riemanniens. Ann. Acad. Sci. Fenn. A. I. 250/35, 1—7 (1958).

STORVICK, D. A.: [1] On meromorphic functions of bounded characteristic. Proc. Amer. math. Soc. 8, 32—38 (1957). — [2] On pseudo-analytic functions. Nagoya Math. J. 12, 131—138 (1957).

STREBEL, K.: [1] On the maximum dilation of quasiconformal mappings. Proc. Amer. math. Soc. 6, 903—909 (1955).

TANAKA, C.: [1] Note on the cluster sets of the meromorphic functions. Proc. Japan Acad. 35, 167—168 (1959).

TEICHMÜLLER, O.: [1] Untersuchungen über konforme und quasikornforme Abbildung. Deutsche Math. 3, 621—678 (1938). — [2] Extremale quasikonforme Abbildungen und quadratische Differentiale. Abh. Preuß. Akad. Wiss., Math.-naturw. Kl. Nr. 22, 1—197 (1939).

TÔKI, Y.: [1] On the behaviour of a meromorphic function in the neighborhood of a transcendental singularity. Proc. Japan Acad. 17, 296—300 (1941). — [2] On the classification of open Riemann surfaces. Osaka Math. J. 4, 191—202 (1952). — [3] On the examples in the classification of open Riemann surfaces (I). Osaka Math. J. 5, 267—280 (1953).

— and K. SHIBATA: [1] On the pseudo-analytic functions. Osaka Math. J. 6, 145—165 (1954).

TSUJI, M.: [1] On the behaviour of an inverse function of a meromorphic function at its transcendental singular point, I, II, III. Proc. Japan Acad. 17, 415—417; 474—475 (1941); 18, 132—139 (1942). — [2] On an extension of Bloch's theorem. Proc. Japan Acad. 18, 170—171 (1942). — [3] On the behaviour of a meromorphic function in the neighborhood of a closed set of capacity zero. Proc. Japan Acad. 18, 213—219 (1942). — [4] On the cluster set of a meromorphic function. Proc. Japan Acad. 19, 60—65 (1943). — [5] On the domain of existence of an implicite function defined by an integral relation $G(x, y) = 0$. Proc. Japan Acad. 19, 235—240 (1943). — [6] On the Riemann surface of an inverse function of a meromorphic function in the neighborhood of a closed set of capacity zero. Proc. Japan Acad. 19, 257—258 (1943). — [7] Theory of meromorphic functions in a neighbourhood of a closed set of capacity zero. Jap. J. Math. 19, 139—154 (1944). — [8] Beurling's theorem on exceptional sets. Tôhoku Math. J. 2, 113—125 (1950). — [9] On meromorphic functions with essential singularities of logarithmic capacity zero. Tôhoku Math. J. 3, 1—6 (1951). — [10] Some theorems on open Riemann surfaces. Nagoya Math. J. 3, 141—145 (1951). — [11] Fundamental theorems in potential theory. J. Math. Soc. Japan 4, 70—95 (1952). — [12] An extension of Bloch's theorem and its applications to normal family. Tôhoku Math. J. **4**, 203—205 (1952). — [13] Theory of meromorphic functions on an open Riemann surface with null boundary. Nagoya Math. J. **6**, 137—150 (1953). — [14] Function of *U*-class and its applications. J. Math. Soc. Japan **7**, 166—176 (1955). — [15] On the cluster set of a meromorphic function. Comment. Math. Univ. St. Pauli **4**, 5—9 (1955). — [16] On a Riemann surface, which is conformally equivalent to a Riemann surface with a finite spherical area. Comment. Math. Univ. St. Pauli **6**, 1—7 (1957). — [17] Potential theory in modern function theory. Maruzen, Tokyo (1959).

Tumura, Y.: [1] Quelques applications de la théorie de M. Ahlfors. Jap. J. Math. **18**, 303—322 (1942). — [2] Recherches sur la distribution des valeurs des fonctions analytiques. Jap. J. Math. **18**, 797—876 (1943).

Ugaeri, T.: [1] On the general potential and capacity. Jap. J. Math. **20**, 37—42 (1950).

Uskila. L.: [1] Über die Existenz der beschränkten automorphen Funktionen. Arkiv för Mat. **1**, 1—11 (1949).

Valiron, G.: [1] Sur les singularités des fonctions holomorphes dans un cercle. C. R. Acad. Sci. (Paris) **198**, 2065—2067 (1934). — [2] Sur les singularités de certaines fonctions holomorphes et de leurs inverses. J. Math. pures appl. **15**, 423—435 (1936). — [3] Remarque sur les domaines complets d'univalence des fonctions entières. Bull. Sci. Math., Ser. II, **63** (1939).

Virtanen, K. I.: [1] Über die Existenz von beschränkten harmonischen Funktionen auf offenen Riemannschen Flächen. Ann. Acad. Sci. Fenn. A. I. **75**, 1—8 (1950).

Weigant, L.: [1] Über die Randwerte meromorpher Funktionen einer Veränderlichen. Comment. Math. Helvet. **22**, 125—149 (1949).

Wittich, H.: [1] Neuere Untersuchungen über eindeutige analytische Funktionen. Berlin: Springer-Verlag 1955.

Wolf, F.: [1] Ein Eindeutigkeitssatz für analytische Funktionen. Math. Ann. **117**, 383 (1940—1941).

Wolkowiskij, I.: [1] Quasikonforme Abbildungen. 1—154. Verlag der Universität Lwow (1954). Russisch.

Yosida, T.: [1] On the behaviour of a pseudo-regular function in a neighbourhood of a closed set of capacity zero. Proc. Japan Acad. **26**, 1—8 (1950). — [2] Theorems on the cluster sets of pseudo-analytic functions. Proc. Japan Acad. **27**, 268—274 (1951).

Yûjôbô, Z.: [1] A theorem on Fuchsian groups. Mathematica Japonicae **1**, 1—2 (1949). — [2] On the Riemann surfaces, no Green function of which exists. Mathematica Japonicae **2**, 61—68 (1951). — [3] An application of Ahlfors' theory of covering surfaces. J. Math. Soc. Japan **4**, 59—61 (1952). — [4] On pseudo-regular functions. Comment. Math. Univ. St. Pauli **1**, 67—80 (1953). — [5] On the quasi-conformal mapping from a simply connected domain on another one. Comment. Math. Univ. St. Pauli **2**, 1—8 (1953). — [6] On absolutely continuous functions of two or more variables in the Tonelli sense and quasi-conformal mappings in the A. Mori sense. Comment. Math. Univ. St. Pauli **4**, 67—92 (1955). — [7] Supplements and corrections to my paper: On absolutely continuous functions of two or more variables in the Tonelli sense and quasi-conformal mappings in the A. Mori sense. Comment. Math. Univ. St. Pauli **5**, 33—36 (1956).

# Subject Index